Coll...

Dinosaurs

Douglas Palmer

HarperCollins*Publishers* Ltd.
77–85 Fulham Palace Road
London
W6 8JB

The Collins website address is:
www.collins.co.uk

Collins is a registered trademark of
HarperCollins*Publishers* Ltd.

First published in 2006

10 09 08 07 06
10 9 8 7 6 5 4 3 2 1

A catalogue record for this book is available from the British Library.

ISBN-10: 0-00-722253-X
ISBN-13: 978-0-00-722253-7

Edited and designed by D & N Publishing, Lambourn Woodlands, Berkshire

Illustrations by James Robins and Steve White (pages 10, 13, 18, 21, 23, 25,
27, 28, 31, 33 only)

Colour reproduction by Digital Imaging, Glasgow
Printed and bound in Italy by Amadeus S.p.A.

CONTENTS

INTRODUCTION

To most people, the word dinosaur brings to mind an enormous animal of awesome strength and ferocity. The name dinosaur means 'terrible lizard', and thanks to a constant deluge of dinosaur images in popular media, these amazing extinct animals have become perhaps the most iconic creatures in history.

What is less often apparent is the extraordinary diversity and variety of the dinosaurs. They were not all vast meat eaters and some resembled birds more closely than the reptiles we know today, with feathers rather than scaly skin. This introduction to dinosaur life portrays a selection of the dinosaurs we know most about, from the earliest ones, such as *Eoraptor* that lived around 228 million years ago, to the latest ones, such as *Triceratops* that became extinct 65 million years ago. We discover what they were like, where they lived, how they behaved, when they were first found and what the most recent discoveries tell us. Also included is information on what their names mean and how to pronounce them.

The popularity of dinosaurs has boomed since the early decades of the 19th century, when fossils of these extinct prehistoric monsters were first discovered and portrayed as once living animals. Yet despite the efforts of dinosaur experts over the last 200 years, we still have just a small sample – some 600 different kinds (genera) – of the total range and diversity of these remarkable animals, which dominated life on land for the best part of 100 million years.

As a result of this, many questions remain unanswered about how they lived, where they came from and how they were related to one another. For example, were the giant meat-eating theropods active hunters that chased down their prey, or were they scavengers who used their size to chase other meat eaters away from their kills? And did the giant plant-eating sauropods continue growing over a long period or did they grow faster than many of the large animals that are alive today? There is still much to be discovered, and who knows, maybe one day you will be the one to answer some of these questions.

WHAT IS A DINOSAUR?

Since the early decades of the 19th century, fossil hunters have dug up hundreds of different kinds of ancient reptiles that we now know as dinosaurs. From tiny bird-sized creatures to the largest animals ever to have lived on land, the dinosaurs were an extraordinary group of animals. Ancient peoples populated their myths and legends with all sorts of strange monsters such as dragons. Today, the role of these creatures has been filled by the dinosaurs – real animals, many of which were just as strange as any fictional beast.

Therizinosaurus, **fantastic fiction or actual dinosaur?**

The dinosaurs first appeared around 230 million years ago and altogether form a special group of reptiles with certain distinct characters that were first recognized in the mid-19th century by the English anatomist Richard Owen. These features are preserved in the structure of their skeletons, especially their hip bones, and they separate dinosaurs from other reptiles both living and extinct. A number of extinct and distant reptile relatives of the dinosaurs, such as the flying pterosaurs and the sea-dwelling ichthyosaurs and plesiosaurs, are often described as if they were dinosaurs, but this is not the case.

Iguanodon, first described by Gideon Mantell

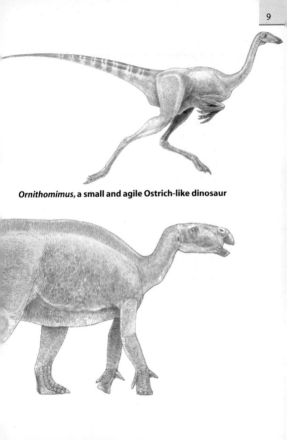

Ornithomimus, a small and agile Ostrich-like dinosaur

A crocodile, a living relative of the dinosaurs

The closest living relatives to the dinosaurs are the birds and crocodiles, with which they share certain features. All three groups share the presence of extra openings in the skull and jaw. With the birds they have in common lightly built bones, the bony structure of their legs, a hinged ankle, and aspects of the skull and jaws. And while the dinosaurs are also similar to crocodiles, they differ from them and other reptiles in the bone structure of their legs, feet and hips. Crocodiles, like most other living reptiles, sprawl with their limbs held out to the side of the body. However, the dinosaurs have their limbs tucked in under the body, a structure that allows them to walk and run more efficiently.

The image and idea of dinosaurs as once living animals has changed enormously since their fossil remains were first discovered and recognized as belonging to a distinct group of extinct reptiles. Even the iconic status of the giant meat eaters like *Tyrannosaurus rex* has been partly displaced by that of small, fast-moving and highly active predators, some of which are known to have been covered with hair-like down and feathers.

Archaeopteryx, **a feathered dinosaur descendant**

DINOSAUR APPEARANCES

Nobody has ever seen a live dinosaur, so our ideas about what exactly they looked like and how they behaved mostly have to come from our interpretation of their fossil record. Using living reptiles as a model for how dinosaurs might have looked led early investigators astray, because in many ways the dinosaurs were very different from crocodiles and lizards.

Once it was realized that dinosaur skeletons were put together in a rather different way from those of any living animals, scientists had to go back to the basics of anatomy to try to reconstruct what the creatures really looked like. As more and better preserved skeletons were found, many surprise discoveries were made. Some dinosaurs – for example, sauropods such as *Diplodocus* – were immense, with massive bodies supported on four pillar-like legs; if anything, they resembled large plant-eating mammals. Other dinosaurs were more like birds in that they supported their body weight on their two hind legs. These dinosaurs could be pretty immense, too, especially some of the meat-eating theropods such as *Allosaurus*.

Accurate reconstructions of the skeletons found have allowed scientists to gain an idea of the size and shape of the 600 or so different kinds of dinosaurs. They can also show how the dinosaurs moved about and, along with the shape of their teeth, indicate how they fed. However, very rarely do fossils tell us anything about other aspects of dinosaurs' behaviour, such as their breeding or social habits, or the fine detail of their appearance, including the texture and colour of their skin. Surprisingly, however, new discoveries are revealing some of these details to us, as we shall see.

For more than 200 years anatomists have realized that the careful study of the way the bones of a skeleton fit and work together – called comparative anatomy – allows the reconstruction of extinct beasts whose body shape may be quite unlike that of any living animal. The reason for this is that all animals with bony

Bodybuilding: *Velociraptor* skeleton and musculature

Stegosaurus, with its distinctive bony back plates

skeletons, from fish to humans (known collectively as the vertebrates), share a common structure. This is still true even when parts of the body perform quite different functions. For instance, the human arm and the wings of birds and bats have a basic structure similar to the front leg of a *Diplodocus*, the arm of *Tyrannosaurus* and the paddle of a whale. By assessing the different proportions and structures of fossil bones with equivalent bones in living animals, their functions in the extinct animal can be predicted.

Detailed study of fossil bones can also indicate the presence of other features such as where muscles attached and how big they were, and where blood vessels and nerves ran. The thickness of a bone's wall and its interior structure indicate how strong it was and whether it could carry a heavy body or was instead adapted for swimming or flight.

Sometimes, fossil skeletons are found that are nearly complete and have their various bones still articulated in their true anatomical positions. Such discoveries have helped confirm what anatomists have predicted, but they have also sometimes led to surprises. Some dinosaur structures are so original and unusual that they cannot easily be predicted by comparative anatomy. For instance, what exactly the back plates of a *Stegosaurus* and the hand claws of a *Therizinosaurus* were used for is still something of a puzzle.

Very occasionally, imprints or mineralized soft tissues are preserved that show that a particular dinosaur had a scaly skin or a feather-like covering. Even internal organs such as the gut and stomach contents have been preserved in some instances, giving important clues as to how the animals lived.

The dome on the skull of *Pachycephalosaurus* could be up to 25cm (10in) thick

DINOSAUR BEHAVIOUR

Much of our understanding of dinosaur behaviour is based on our knowledge of their body form and feeding habits – whether they were plant eaters (herbivores) or meat eaters (carnivores). Knowledge of how different kinds of living plant eaters interact with their meat-eating predators can give insights into how the dinosaurs interacted. Like herbivorous mammals today, many plant-eating dinosaurs tended to have good senses of sight, smell and hearing. To escape their predators, some were slimly built and fleet of foot. Others had heavy defensive armour and yet others sought safety in numbers. The meat eaters were well equipped with offensive weapons such as large, sharp teeth and claws, and were either fast on their feet, large and stealthy or good ambush hunters.

The study of dinosaur feeding behaviour was originally based on that of large living reptiles such as crocodiles. However, this was somewhat misleading as crocodiles, with their relatively small limbs, largely live in water and can survive on infrequent feeds. Relating other aspects of crocodile behaviour to dinosaurs proved equally mistaken. Crocodiles, for example, belong to the group of animals known as ectotherms, which rely on the sun's rays to heat their bodies. In contrast, endotherms (which include mammals) use energy from frequent supplies of nutritious food and complex blood systems to maintain a relatively high body temperature. It was once thought that all dinosaurs were crocodile-like ectotherms, but we now know from several different lines of evidence, such as their bone structure, that some dinosaurs may have been endotherms. Consequently, their levels of activity and feeding were probably more like those of mammals than ectothermic reptiles living today.

We can now envisage many land-living predatory dinosaurs as active hunters that could run at speeds of up to 40kph (25mph) to catch their prey. Some predators may even have worked together in groups to kill prey larger than themselves, and some were cannibals. Many of the larger herbivores lived in herds so that their many combined eyes, ears and nostrils could provide early warning of approaching predators. But how do we know what dinosaurs ate?

Pelecanimimus had a relatively large brain and unusually small teeth

DINOSAUR DIETS

The most important clues as to what dinosaurs ate are provided by their teeth. Typically, the teeth of dinosaurs (like those of other reptiles) were constantly replaced from within the jaw throughout the animal's life.

The types of food an animal eats determine the type of teeth it requires. Meat eaters need sharp, dagger-like and blade-shaped teeth for killing their prey and tearing pieces of meat from the carcass. There are, however, further differences within this group

Meat-eaters: a *Tyrannosaurus rex* skull and its cutting teeth

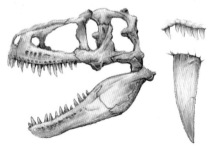

depending upon whether the prey consists of insects, fish, small mammals or large, well-defended, aggressive plant eaters or other meat eaters.

Small insect-eating dinosaurs, for example, needed only small, sharp, conical teeth that were strong enough to crunch the insects up. In contrast, large predatory dinosaurs that required lots of meat had large teeth that were powerful and sharp enough to kill their prey and then remove as much flesh as quickly as possible before other predators and scavengers were attracted to the kill. Some of these teeth had sophisticated cutting edges with serrations like those on a steak knife.

Plant-eating dinosaurs displayed an even greater variety in the form of their teeth depending upon the kind of plant material they ate. Many of the giant sauropods had strange peg-like teeth that probably worked like rakes, stripping large amounts of foliage from branches as the animal's long neck swept from side to side and up and down. Others herbivores ate much tougher fibrous plants and had large, strong, leaf-shaped teeth that were gradually worn away and eventually replaced.

Since flowering plants did not evolve until Cretaceous times around 100 million years ago, the food available

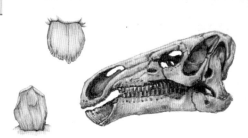

Plant-eater: *Iguanodon* and details of its grinding teeth

to herbivorous dinosaurs included more basic groups of plants such as ferns, conifers, cycads, horsetails and ginkgos. Many of these plants were difficult to digest and had low nutritional value. Like many herbivorous mammals today, the plant-eating dinosaurs had large stomachs and had to eat constantly to obtain enough nutrition from their diet. As an aid to digestion, some of them swallowed stones that then acted as a gastric mill along with their stomach flora of digestive bacteria. But large size slows an animal down, so how did the giant herbivores protect themselves from predators?

EGGS, NESTS AND BABIES

Fossil eggs, now known to be those of dinosaurs, were first found in France and England in the mid-19th century. But it was not until the amazing discoveries of the Roy Chapman Andrews' expedition to Mongolia in the 1920s that we really began to get an idea of how dinosaurs reproduced in bird-like ways and how some of them looked after their babies.

Chapman and his team from the American Museum of Natural History in New York found fossil eggs and

A nesting *Psittacosaurus* dinosaur with babies

hatchlings clustered together in and around mud mounds. They even thought that they had found the remains of a dinosaur egg thief called *Oviraptor* in the act of stealing from a *Protoceratop*'s nest. However, we now know that it is more likely that the *Oviraptor* died defending its own nest.

Dinosaur eggs ranged in size and shape from tiny round eggs to long oval-shaped ones more than 50cm (20in) long and around 4.0l (8.5pts) in volume. These were, however, small compared to the biggest bird egg, which was laid by the extinct Elephant Bird (*Aepyornis*) and measured more than 1m (3ft) in circumference and 7.3l (15.4pts) in volume. Large eggs need thick shells but these also have to be porous to allow the foetus to 'breathe'. Consequently, dinosaur eggs could not be too large. Like modern-day turtles, many dinosaurs laid large numbers of eggs because relatively few of the babies would survive into adulthood.

In the 1970s, finds from the so-called Egg Mountain site in Montana, USA, revealed what may have been a shared nesting ground and hatchery for the duck-billed dinosaur *Maiasaura*. Several nests up to 2m (6.6ft) wide have been found, made of layers of plant material and mud topped by a hollow in which up to 12 eggs were laid. The 9m-long (30ft) mother was far too big to have sat on the eggs (see pages 162–3) but probably would have fed her babies, which could not have looked after themselves to begin with.

In contrast, hatchlings of the small theropod *Troodon*, which also nested around Egg Mountain, were much more precocious and probably began to feed themselves very early on. Surprisingly, there is evidence that even some of the biggest plant-eating sauropods stayed around their nests for some time, presumably to try to protect the eggs and babies from predators.

A *Mussaurus* hatchling. Some of the skeletons of this species could fit into the palm of a hand

THE ARMS RACE

There are a number of reasons why animals, and even humans, fight one another. Disputes over food, territory and mates commonly lead to conflict, and dinosaurs were no exception to the rule. However, our understanding of dinosaur disputes is biased by what information we can recover from the fossil record and what can be inferred from general principles of animal behaviour.

Obviously, the main arena of conflict results from the predator–prey relationship between meat eaters and plant eaters. Predators have to be appropriately armed to catch and kill their prey and, as we have seen, this is primarily a matter of being able to detect the prey, surprise it and then catch it with claws or teeth. Inevitably, eggs, babies and juvenile dinosaurs would have been most vulnerable to predation. From the prey's point of view, the matter is largely one of flight or fight – in other words, being able to run away or defend itself.

Some of the small plant-eating dinosaurs opted for the flight mechanism and were fleet of foot. The two-legged ornithomimids, for example, may have been able to run at speeds of up to 40kph (25mph). At the other end of the scale, giant sauropods, once grown up, were so big they would have been invulnerable. Medium-sized but heavy, slow plant eaters such as the ceratopsians, stegosaurs and ankylosaurs evolved a variety of types of armour and defensive weaponry. The ceratopsians are characterized by their helmet-like skulls with prominent rhino-like horns. The stegosaurs had plates and spikes along their back and tail, while the ankylosaurs had tough bony plates embedded in their skin as well as bone-crunching tail clubs. The same weaponry was probably also used by males when fighting one another in territorial disputes or over females. Some of the defensive weaponry that is not as structurally robust as

Repenomamus **with a baby** *Psittacosaurus*

it looks, such as the neck flanges and frills of the ceratopsians, may also have had other functions, such as for male display, signalling and species recognition.

Medium-sized plant eaters that were neither fast movers nor well armoured developed other means of defending themselves against predators. The hadrosaurs, for example, were equipped with a variety of strange bony structures on their skulls, some of which may have been used to amplify calls to other members of the herd to warm against a nearby predator.

And, finally, we now know that dinosaurs did not have it all their own way. Recently, the fossil of a badger-sized mammal called *Repenomamus* has been found with the remains of a baby psittacosaur in its stomach cavity.

Brachiosaurus, **quite possibly too large for most predators**

TRACKS AND TRAILS

When dinosaur footprints were first seriously studied they presented a considerable puzzle. Footprints and tracks found in the early 19th century in Massachusetts, USA, were initially seen as evidence for the Old Testament Flood. But by the 1830s the Reverend Professor Hitchcock of Amherst College had developed the new science of palaeoichnology, or the study of fossil footprints. He thought that the abundant three-toed footprints that occurred in strata dating from the Late Triassic (210 million years ago) and Jurassic (200–145 million years ago) must have been made by birds, some of which had to be much larger than any species living today. We now know, however, that these particular prints were made by two-legged (bipedal) dinosaurs.

The modern study of dinosaur prints can tell us a great deal about the animals' behaviour. For instance, from print size, the distance between prints and estimates of the leg length of the animals that made them, it is possible to calculate the speed at which the animal was moving. Some small, lightly built bipedal dinosaurs seem to have been capable of running at speeds of up to 40kph (25mph).

The largest quadrupedal sauropods were so massive that they were capable of only a fast saunter at around 20kph (12mph).

Sets of trackways made by the same kind of animal have been found, showing that certain dinosaurs moved around in herd-like groups, especially plant eaters such as the iguanodontids and hadrosaurs. Presumably this behaviour was for protection. In contrast, many theropod trackways tend to be solitary, showing that they were lone hunters, although not all behaved in this way – some theropods are also known to have grouped together.

The problem with many footprints and tracks is that it can be difficult or impossible to link the prints with a particular dinosaur, as very few animals literally dropped dead in their tracks. Instead, interpretation has to come from our understanding of the structure of the limbs, ankles and number of toes possessed by different dinosaurs. Tracks have helped show that some dinosaurs such as the iguanodontids, which were originally thought to have been bipedal, normally moved on all fours even though their fore limbs were smaller than their hind limbs.

THE DINOSAUR ERA

The dinosaurs first appeared around 230 million years ago in Late Triassic times. Over the following 165 million years, through the Jurassic and Cretaceous periods, they increased enormously in both abundance and variety, only to die out abruptly 65 million years ago at the end of the Cretaceous. But in a way the dinosaurs are still with us, only they are now all feathered and we know them as birds.

The earliest dinosaurs included members of both major groups: the saurischians and ornithischians. Many of the oldest dinosaur fossils have been found in South America, which at the time was connected to Africa, India, Antarctica and North America as part of the much larger supercontinent of Pangaea. These fossils have sufficient features in common to suggest that they were derived from an even earlier and as yet unknown common ancestor in Middle Triassic times.

By the end of the Triassic period many different groups of dinosaurs had evolved as well as the flying pterosaur reptiles. As global climates became drier, conifers became more abundant and plant-eating dinosaurs became more common than other reptilian plant eaters. By the Jurassic period the main kinds of dinosaurs had appeared, but alongside them were crocodile-like reptiles, turtles and early mammals. The dinosaurs had also spread across Pangaea from the southern hemisphere into the northern hemisphere and most continents to become a worldwide success.

The Jurassic saw the rise of the first really gigantic plant-eating sauropods and some of the massive predatory theropods such as the allosaurids. The earliest fossils of primitive birds (*Archaeopteryx*) date from the end of this time period, showing that their ancestry and origin from dromaeosaur (raptor) dinosaurs had occurred earlier. Unfortunately, no fossil record of this important event has been recovered, and much of our information on these birds is from later Cretaceous fossils.

By the end of Cretaceous times, 65 million years ago, dinosaurs were well established across the earth from pole to pole. However, they all suddenly died out in what is called the end-Cretaceous extinction event, along with many other kinds of animals and plants both on land and in the sea. The event coincides with the collision of an 11km-wide (7-mile) asteroid-like rock with the earth. What is surprising about this is that many other kinds of reptiles such as the crocodiles and turtles survived, along with the mammals and those dinosaur descendants, the birds.

FOSSIL FORMATION

It can be surprisingly difficult for the remains of any land-based animal to become fossilized, which is why we have relatively few dinosaur and human-related fossils. When most animals die on land, their corpses are scavenged and are often totally destroyed unless they contain large bones or other indigestible structures such as teeth. Prolonged exposure to weathering and erosion by wind and water further scatters and wears away the remaining bones. The preservation of an

Bodies are food for scavengers

Exposed to the elements, the flesh decays

entire dinosaur skeleton would have required quick burial below new layers of sediment, a situation that occurs only in certain circumstances.

Fossil hunters are able to spot the remains only after the sediment layers (strata) have been brought back up to the surface and re-exposed through earth movements or deep erosion. Experts have now learned to search out the right kind of strata that were originally laid down in the sort of environments occupied by dinosaurs and where their remains might have been buried. The deposits of rivers, lakes and near-shore deltas have proved good prospects, along with the deposits of more arid environments where footprints can be preserved.

Most dinosaurs are known only from incomplete and fragmentary skeletal remains. Commonly, the skull is missing, along with the hands and feet. Of the 600 or so different kinds of dinosaurs known so far, most are represented by just a single species. We have probably found only the tip of the proverbial iceberg. Even when a fossil has been found, identification is not necessarily certain as it depends on whether the most important characteristics have been preserved. Sometimes specimens of the same kind of dinosaur have been given different names because they have been found in different countries, or because different bits of the same animal have been found in different places.

Resurrection in rock

THE FIRST FINDS

The first dinosaur fossils were probably found in China many hundreds of years ago, when they were thought to be dragon bones. In modern times it was the early 19th-century discovery of some puzzling fossil bones in the south of England that really started the dinosaur story. In the 1820s, a large 25cm-long (10in) jawbone set with sharp, blade-shaped teeth excavated from Jurassic strata in Oxfordshire was described by William Buckland as *Megalosaurus*, a new kind of giant and extinct reptile. Around the same time, Gideon Mantell was trying to reconstruct another large fossil reptile that he called *Iguanodon* from a jumble of bones found in Sussex.

But both Buckland and Mantell were trounced by anatomist Richard Owen, who was the first to distinguish and name the Dinosauria (meaning 'terrible lizard') as an extinct group of reptiles in 1842. Owen realized that both *Megalosaurus* and *Iguanodon* had features that distinguished them as dinosaurs. For the reopening of Crystal Palace in south London in 1853, Owen and the sculptor Waterhouse Hawkins built the first life-sized dinosaur models, which still exist at the site today. They saw

Megalosaurus and *Iguanodon* as huge, lumbering, four-legged and rather elephantine beasts with massive tails.

By the end of the 19th century the image of the dinosaurs had been transformed by the discovery of much more complete fossil skeletons, especially in North America. It was realized that some dinosaurs had moved around on their large, powerful hind legs, and it was discovered that the dinosaurs could be divided into two main groups: the lizard-hipped saurischians; and the bird-hipped ornithischians. Some of the first giant sauropod skeletons had been found, as well as one of the best known of the giant carnivorous theropods – *Allosaurus*.

Since then, scientific understanding of the dinosaurs has been revolutionized by new techniques of investigation and new fossil discoveries. From the 1920s, new dinosaur-rich locations have been found in many parts of the world, from Mongolia to Argentina and China. Over the last few decades, Cretaceous localities in China have revealed exceptionally well-preserved and complete skeletons, incredibly some of which even retain indications of the original soft tissues. These finds have caused another revolution in our understanding of dinosaur evolution.

DINOSAUR NAMES

Scientists divide dinosaurs and all other organisms into groups of related forms. Traditionally, the groups have been placed in classes, orders, families and so on, but a new system of classification called cladistics has supplemented these names. Here, branches (clades) are defined by characters that evolved at, or immediately before, their origin. For instance, the bird clade, called Aves, is defined by the possession of wings and primary flight feathers.

Each group has a number of unique designated characteristics that should be found in each of its members. For instance, the ankylosaurs were just one of several groups of four-legged plant-eating dinosaurs but they can be distinguished from the others by rows of bony plates embedded in the skin of their backs, a distinctive bony skull form and small denticulate teeth. The ankylosaurs can be grouped together with the stegosaurs (characterized by their prominent bony back plates) as thyreophorans because of the extensive nature of the armour they had on their backs. The thyreophorans include all dinosaurs that are more closely related to *Ankylosaurus* than to *Triceratops*, which belongs in another related group called the marginocephalians, consisting of bone-headed and horned plant-eating dinosaurs.

Both thyreophorans and marginocephalians had distinctive bird-like hip bones that characterize them as belonging to the ornithischians, one of the two main dinosaur groupings. The animals in this group were all plant eaters and most moved around on all four legs. They include such well-known groups as the ankylosaurs, stegosaurs, ceratopsians, and ornithopods (with birdlike feet).

The other main group of dinosaurs is the saurischians, which had a hip structure more like that of lizards. A major subdivision within this group separates the

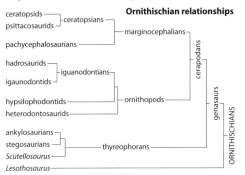

Ornithischian relationships

plant-eating, four-legged sauropods and the meat-eating, two-legged theropods. In recent decades the classification of the theropods has been transformed by new evidence indicating that the birds originated within a theropod group known as the maniraptorans.

The theropods group contains the most subdivisions of any dinosaur group, which reflects the considerable range in form and size of its members. These include some of the largest land-dwelling animals, the allosaurs and tyrannosaurs, which reached lengths of 14m (45ft) and weights of up to 6 tonnes (5.9 tons). At the other end of the scale the theropods include tiny dinosaurs such as *Microraptor*,

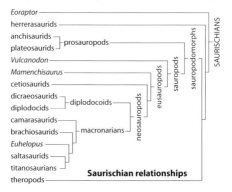

Saurischian relationships

which had a body length of around 47cm (1ft 7in). Aves also belongs to the theropods and numbers *Archaeopteryx* among its dinosaur members; it also includes all of the modern-day birds.

A final group of small to medium-sized meat-eating bipedal saurischians is still difficult to categorize. These include *Herrerasaurus*, *Eoraptor* and *Staurikosaurus*, which have been found in Late Triassic rocks, especially in South America. Although it is still not clear what their relationships are to the main groups of saurischians, these dinosaurs were certainly more like the theropods than the sauropodomorphs.

Theropod relationships

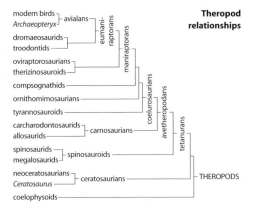

ORIGINS AND EVOLUTION

The fossil record tells us that from around 230 million years ago, in the Late Triassic Period, the dinosaurs evolved into an amazing diversity of forms, these represented by some 600 separate genera. They became increasingly successful in most land environments across the globe from Alaska to Antarctica, and they thrived for more than 160 million years until the end of Cretaceous times 65 million years ago.

The extinction event that marked the end of the Cretaceous saw the eclipse of all the remaining dinosaurs except for the birds. Interestingly, we now have fossil evidence showing that in Cretaceous times some mammals had evolved to sufficient size to eat young dinosaurs. It is possible that if there had not been an extinction event the adaptational advantages of the mammals may have eventually caused the dinosaurs to decline.

Altogether the dinosaurs belong within a bigger grouping of reptiles called the archosaurs, which includes today's crocodiles as well as the extinct flying pterosaurs and other extinct Triassic reptiles.

The discovery and study of Triassic archosaurs has provided our present understanding of the origin of the dinosaurs.

The basic dinosaur groupings of ornithischians and saurischians (see pages 36–39) represent a very early split in the evolution of the animals, which probably occurred in Middle Triassic times. However, some of the earliest dinosaurs, such as *Eoraptor* and *Herrerasaurus*, are basal saurischians. In addition, there are some small reptiles of Middle Triassic age, such as *Marasuchus* and *Lagosuchus*, which, like the earliest dinosaurs, were bipedal – that is, they walked and possibly ran on their slim hind legs with the body weight counterbalanced by a long tail. This adaptation was a major event in the evolution of land animals, as almost all the earlier backboned animals moved around on four legs. The bony structure of the hips and legs that allowed these reptiles to become bipedal is very similar to that seen in fossils of early theropod and ornithischian dinosaurs. This suggests that the origin of the dinosaurs may therefore lie within these bipedal archosaur reptiles, which are called dinosauromorphs. No doubt our understanding of dinosaur evolution will become clearer as more fossils are uncovered.

EORAPTOR
EE-oh-RAPP-tor

'Dawn thief'

Size Up to 1m (3ft), 10kg (20lb).

Appearance This small reptile carried itself on two muscular hind legs, the weight of its body, neck and small 12cm-long (5in) head counterbalanced by a long, stiff tail. The bony structure of the hip, legs and ankle were like those of the saurischian dinosaurs. However, features of the skull, hand and wrist were not fully theropodan, suggesting that *Eoraptor* was either a very close relative of the theropods or a very primitive one.

Behaviour Several features show that this was a meat-eating predator. The size and structure of the legs, ankle and foot indicate that it was capable of fast movement for hunting and chasing its prey. In addition, the structure of the long-fingered hands, with their curved claws, shows that it was well adapted for catching and grasping prey.

Food *Eoraptor*'s jaws were full of curved, serrated teeth like those of many theropods, but its jaw hinge was different. It also had other, more leaf-shaped, teeth typical of many plant eaters, so it may have had a mixed diet of both plant and animal food.

Where did it live? Two nearly complete skeletons belonging to a single species were found in Late Triassic strata in a region known as the Valley of the Moon in northwestern Argentina in the early 1990s and were first described in 1993.

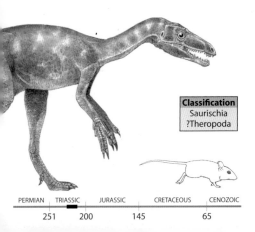

Classification
Saurischia
?Theropoda

PERMIAN	TRIASSIC	JURASSIC	CRETACEOUS	CENOZOIC
251	200	145	65	

HERRERASAURUS

'Herrera's lizard'

huh-RARE-uh-SORE-us

Size 3–4.5m (10–15ft), 300kg (660lb).

Appearance This South American carnivore is one of the best-known early theropod-like saurischian dinosaurs. Although not very big, *Herrerasaurus* was a top predator and one of the larger two-legged animals of its time. Most of the skeleton is known and shows a mixture of primitive and surprisingly

advanced features. For instance, it had some unusually long canine-like teeth. Like *Eoraptor*, which lived nearby, *Herrerasaurus* had relatively small arms but powerful hands with three fingers that ended in long, curved claws.

Behaviour *Herrerasaurus* was highly active and could run and jump using its long, stiff tail as a counterbalance. The hands were well adapted for grabbing prey and its flexible jaw improved the grip and bite on its victim. There is no proof of any nesting or other social behaviour in these early dinosaurs.

Food Evidence for this dinosaur's choice of prey comes from the remains of pig-like plant-eating rhynchosaur reptiles found inside the stomach of a *Herrerasaurus* fossil. However, bite marks on a *Herrerasaurus* skull show that it was also preyed upon by other meat eaters, perhaps even members of its own kind.

Where did it live? A single species was first described in 1963 from a fossil found in Late Triassic strata in northwestern Argentina. Several partial skeletons and a complete skull are now known.

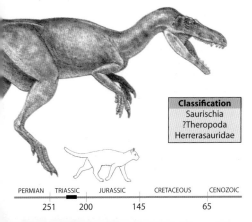

Classification
Saurischia
?Theropoda
Herrerasauridae

PERMIAN	TRIASSIC	JURASSIC	CRETACEOUS	CENOZOIC
251	200	145		65

DILOPHOSAURUS

'Two-ridge lizard'

die-LOH-foh-SORE-us

Size Up to 7m (23ft), 500kg (1,100lb).

Appearance One of the earliest of the large meat-eating dinosaurs, *Dilophosaurus* was the largest predator of its time. It can be recognized by the two distinctive V-shaped bony crests that extended from its nostrils back over the crest of the head. Its powerful legs ended in long clawed toes.

Behaviour Its toe claws would have given *Dilophosaurus* a good grip when running and may also have been used to help pin down its prey. The first finger of its powerful four-fingered hands was like a grasping thumb and the large blade-like teeth in its jaws were used to tear the prey apart. The bony head crest was too fragile to have offered any protection to the skull and it is likely that it was for sexual display or some other form of signalling between individuals.

The same kind of double-vaned head crests have been found on a similar Chinese dinosaur, although it is not clear whether the two are in fact related.

Food *Dilophosaurus* probably hunted small plant-eating dinosaurs such as the prosauropod *Ammosaurus* and the ornithischian *Scutellosaurus* (see page 134).

Where did it live? The first and only known species was first described as a megalosaur in 1954 and only recognized as this new genus in 1970. The remains of some six individuals are now known from Early Jurassic strata in the southwestern USA and perhaps also China.

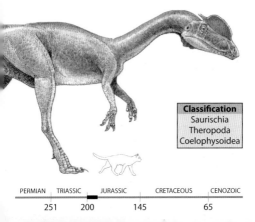

| **Classification** |
| Saurischia |
| Theropoda |
| Coelophysoidea |

PERMIAN	TRIASSIC	JURASSIC	CRETACEOUS	CENOZOIC
251	200	145	65	

COELOPHYSIS

SEEL-oh-FIE-sis

'Hollow form'

Size Up to 3m (10ft), 30kg (66lb).

Appearance *Coelophysis* was a small, fast-running and agile theropod with a slender build and a long neck that was counterbalanced by an even longer stiff tail. Thanks to the discovery in 1947 of around a thousand *Coelophysis* skeletons all jumbled together at Ghost Ranch in New Mexico, USA, we know a great deal about this small but highly active predator. Its legs were long and slender, while the arms were fairly short and ended in

three long fingers with narrow claws, a fourth finger without a claw and a fifth residual finger.

Behaviour The fact that so many skeletons have been found in one place suggests that a herd of these dinosaurs was overcome by a flash flood and swept away, to be buried together. They may have herded together during the breeding season, which carried its own risks of fighting and cannibalism.

Food This very active predator seems to have eaten almost anything it could catch. Juvenile *Coelophysis* skeletons have been found in the stomach cavity of some adults; at first these were thought to be embryos, but the bones are too well developed and it is more likely that the adults were cannibalistic.

Where did it live? *Coelophysis* was first discovered in Late Triassic strata in New Mexico, southwestern USA, and described by Edward Drinker Cope in the late 1880s. This single species genera is now also known from Arizona.

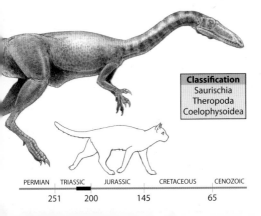

Classification
Saurischia
Theropoda
Coelophysoidea

PERMIAN	TRIASSIC	JURASSIC	CRETACEOUS	CENOZOIC
251	200	145		65

CERATOSAURUS
Seh-RA-toh-SORE-us

'Horned lizard'

Size Up to 6m (20ft), 815kg (1,800lb).

Appearance This medium-sized, two-legged meat eater can be identified by the combination of a single nose horn and pair of smaller horns above its eyes. Similar head ornaments found in other related dinosaurs allows them to be grouped together as ceratosaurians. They survived for at least 155 million

years and had an almost global distribution, being found on all the continents except Australia and Antarctica. Overall they ranged in size from small, lightly built animals such as the 1.5m-long (5ft) *Noasaurus* right up to the 7.5m-long (25ft) *Carnotaurus* (see page 52).

Behaviour The head structures in *Ceratosaurus* were not particularly robust and so would not have been any use for defence. It is more likely that they were for display and may have distinguished males from females.

Food *Ceratosaurus* was probably a lone hunter that preyed upon relatively small plant-eating ornithopod dinosaurs and other reptiles.

Where did it live? It was first described from remains in Late Jurassic strata in western USA by Othniel Marsh in 1884. Today, fossils of some five individuals are known from here, along with those of *Allosaurus*, an even bigger predator (see page 62). Two other possible species from the same region were described in 2000.

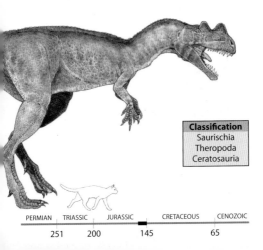

| **Classification** |
| Saurischia |
| Theropoda |
| Ceratosauria |

PERMIAN	TRIASSIC	JURASSIC	CRETACEOUS	CENOZOIC
251	200	145	65	

CARNOTAURUS

'Meat [-eating] bull'

KAR-noh-TORE-us

Size Up to 7.5m (25ft), 1 tonne (2,200lb).

Appearance This powerful theropod had muscular legs with long shins and toes like the tyrannosaurs, as well as very short arms ending in four fingers. It also had massive breast bones, and the skull was deep with a short snout and horns projecting out above the eye sockets. Unusually, the eyes faced forward and may well have provided good depth perception. The lower jaw had a flexible hinge

and the mouth was filled with relatively small but sharp teeth.

Behaviour With legs that were adapted for running, *Carnotaurus* was an agile hunter. It is not yet known whether the skull horns were used for male display and fighting, or whether both sexes had them for defence.

Food The skull was relatively light and flexible, so it is unlikely that *Carnotaurus* would have tackled large prey. It may, however, have used its superior distance judgement to specialize in catching small, agile prey.

Where did it live? The isolation of South America from the northern continents 144 million years ago meant that the dinosaurs living there developed into somewhat different forms from those elsewhere. *Carnotaurus* was first described in 1985 by Argentinian dinosaur expert J. F. Bonaparte from a sole complete skeleton of the only species found in Late Cretaceous strata of Patagonia.

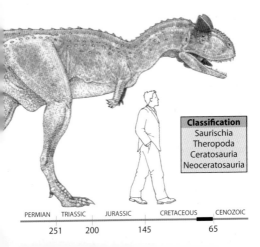

| **Classification** |
| Saurischia |
| Theropoda |
| Ceratosauria |
| Neoceratosauria |

PERMIAN	TRIASSIC	JURASSIC	CRETACEOUS	CENOZOIC
251	200	145	65	

MEGALOSAURUS

MEGA-loh-SORE-us

'Great lizard'

Size Up to 8m (26ft), 1 tonne (2,200lb).

Appearance Historically famous, *Megalosaurus* was one of the first giant, extinct land-living reptiles described – in a work completed in 1824 by William Buckland of Oxford University. It was also one of the first animals recognized as a dinosaur, in 1842 by English anatomist Richard Owen. To begin with, very few bones were known except for a jawbone filled with long, curved, blade-like, serrated teeth that

showed it to have been a ferocious predator. Owen reconstructed the animal as a great, lumbering four-legged elephantine beast. We now know that a strong neck supported its massive head and that this was balanced by a heavily muscled and stiff tail.

Behaviour *Megalosaurus* was a powerful two-legged hunter. It had features in common with other dinosaurs such as *Baryonyx* (see page 56), and so these are grouped together as spinosauroids within the early tetanurans, which is made up of all dinosaurs that share a more recent common ancestor with birds than with the ceratosaurians.

Food The form of its teeth shows that *Megalosaurus* was a meat-eating predator.

Where did it live? *Megalosaurus* was first found in the Stonesfield Slate of Middle Jurassic age in Oxfordshire, England. Only a single species is well known and it is still not thoroughly understood as no complete skeleton has been found. Over the decades, some 14 different species have erroneously been placed in the genus.

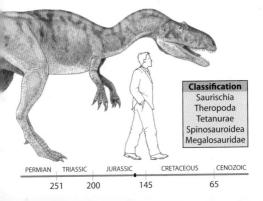

| **Classification** |
| Saurischia |
| Theropoda |
| Tetanurae |
| Spinosauroidea |
| Megalosauridae |

| PERMIAN | TRIASSIC | JURASSIC | CRETACEOUS | CENOZOIC |
| 251 | 200 | 145 | 65 | |

BARYONYX   *'Heavy claw'*
ba-ree-ON-iks

Size Up to 12m (39ft), 2 tonnes (1.9 tons).

Appearance This large spinosaur is of particular interest because of its shape, which differed from that of most other meat eaters, and because of its specialized diet. It had a distinctive long, narrow skull with the nostrils set back some 10cm (4in) instead of being right at the end of the snout. The lower jaw held twice as many teeth as the upper jaw and the front teeth were bigger than the rest. The teeth were also wider with finer serrations than the typical 'steak-knife' teeth of other meat-eating dinosaurs. The arms were short but powerful, and the hands had three fingers ending in curved, sickle-shaped claws 30cm (1ft) long.

Behaviour These features of the teeth and arms allowed *Baryonyx* to hunt and grab slippery, wriggling fish and hold them firmly in its mouth before swallowing them whole.

The claws would also have been very effective for defence against other predators.

Food Fish remains and *Iguanodon* bones were found in the stomach region when the first *Baryonyx* find was excavated, suggesting that it was a fish eater and perhaps a scavenger.

Where did it live? The region of southern England and Spain where *Baryonyx* lived in Early Cretaceous times was a swampy subtropical coastal plain with wide deltas and rivers full of fish, some of which grew to 3m (10ft) long. *Baryonyx* was first discovered in 1983 and the single known species was described by English dinosaur specialists Alan Charig and Angela Milner.

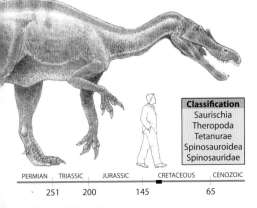

Classification
Saurischia
Theropoda
Tetanurae
Spinosauroidea
Spinosauridae

PERMIAN	TRIASSIC	JURASSIC	CRETACEOUS	CENOZOIC
251	200	145	65	

SPINOSAURUS

'Spine lizard'

SPY-noh-SORE-us

Size Up to 14m (45ft), 5.1 tonnes (5 tons).

Appearance A series of tall bony spines, up to 1.5m (5ft) high and covered with skin, formed a 'sail' along the backbone of this dinosaur, distinguishing it as a spinosaur. Two massive muscular legs supported the body at the hips, and a stiff tail counterbalanced the chest, neck and crocodile-like head. The jaws were also armed with straight, cone-shaped, crocodile-like teeth.

Behaviour The sail of *Spinosaurus* may have helped the animal warm up quicker than other hunting dinosaurs – if it was turned sideways to the sun it would have acted like a solar panel. With its large surface area it could also have helped this relatively large animal cool down in the heat of the day. However, the sail may also have been a display feature for attracting mates.

Food The crocodile-like skull and teeth suggest that this predator ate fish and other kinds of meat.

Where did it live? *Spinosaurus* was originally collected by the German palaeontologist Ernst Stromer von Reichenbach from Late Cretaceous strata of Egypt in 1912 and described in 1915, but the specimens were destroyed in World War II. A second species was described from Morocco in 1996 from a fragmentary skeleton.

Classification
Saurischia
Theropoda
Tetanurae
Spinosauroidea
Spinosauridae

PERMIAN	TRIASSIC	JURASSIC		CRETACEOUS	CENOZOIC
251	200	145		65	

CRYOLOPHOSAURUS
CRY-oh-LOAF-oh-SORE-us

'Frozen crested lizard'

Size Up to 7.5m (24ft), 815kg (1,800lb).

Appearance The feature that distinguishes the remains of this theropod is the strange backward-sweeping crests of bone along the top of the skull above the eyes. A vertical crest with a furrowed surface extends across the width of the skull between the eyes. Although the single fossil find is very incomplete, *Cryolophosaurus* has been declared the oldest known tetanuran.

It shows that tetanurans and ceratosaurs had diverged by Early Jurassic times.

Behaviour The skull crest, like those of other crested theropods such as *Dilophosaurus* (see page 46), was probably used for courtship display and/or signalling to other members of the species.

Food The serrated and dagger-shaped teeth show that *Cryolophosaurus* was a carnivore like most theropods and probably an active predator.

Where did it live? It was first discovered in 1990 by geologist David Elliott in an Early Jurassic rock outcrop high on Mt Kirkpatrick in Antarctica. The single known species was described in 1994. Its presence here along with the remains of a prosauropod dinosaur and a flying pterosaur reptile, shows that Antarctica was still connected to the southern Gondwanan continents and part of the supercontinent of Pangaea at this time. Although the continent was close to the South Pole, and it had dark winters, it was warm enough to be inhabited by an abundance of plants and animals.

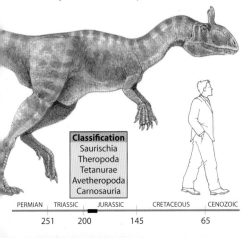

Classification
Saurischia
Theropoda
Tetanurae
Avetheropoda
Carnosauria

PERMIAN	TRIASSIC	JURASSIC	CRETACEOUS	CENOZOIC
251	200	145		65

ALLOSAURUS
al-loh-SORE-us

'Different lizard'

Size Up to 14m (45ft), 3 tonnes (3 tons).

Appearance *Allosaurus* was one of the most impressive meat eaters ever to have lived. A giant on muscular legs, balanced by a stiff tail, it had a huge head, a powerful neck and jaws, and sharp, curved, serrated teeth. Its arms were small but strong, with the hands' three fingers ending in sharp, curved claws. The skull carried distinctive bony ridges from the eye to the snout; ridges with pointed crests probably belonged to males and rounded ones to females.

Behaviour Being so big, *Allosaurus* could not run fast and could catch small dinosaurs only by ambushing them. However, it was fast enough to catch giant sauropods and may have hunted in small groups to isolate vulnerable individuals from their herds.

Food *Allosaurus* preyed on other dinosaurs, especially slow-moving plant eaters such as the giant sauropods, smaller plant eaters such as *Stegosaurus* and almost anything else it could catch. It may also have used its size to drive other predators from their kills in order to scavenge the corpses.

Where did it live? It was first found in Colorado, USA, and described by Othniel Marsh in 1877. Most fossils have been found in Late Jurassic strata in western USA. The remains of some 60 individuals are now known, including some complete skulls and skeletons. Most of these belong to a single species, although there may be another American species and an African one from Tanzania. Bones found in Portugal may also belong to the *Allosaurus* genus.

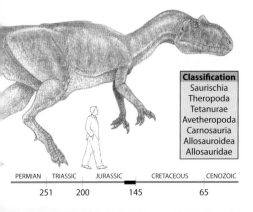

Classification
Saurischia
Theropoda
Tetanurae
Avetheropoda
Carnosauria
Allosauroidea
Allosauridae

PERMIAN	TRIASSIC	JURASSIC	CRETACEOUS	CENOZOIC
251	200	145	65	

CARCHARODONTOSAURUS 'Shark-toothed lizard'
kar-CARR-oh-DONT-oh-SORE-us

Size Up to 14m (45ft), 4 tonnes (4 tons).

Appearance Along with its contemporary relative *Giganotosaurus* from South America and the more distantly related and younger tyrannosaurs from North America, *Carcharodontosaurus* was one of the largest predators ever to have lived on land. Its massive body, powerful neck and enormous head

were carried by the powerful hind legs and counterbalanced by a massive stiff, muscular tail. The skull grew to 2m (6ft) long but its bones were reduced to a framework to cut down weight.

Behaviour Although the arms were relatively small, they ended in three clawed fingers that were perfectly good for grasping their prey. The wide gape of the jaws and the strength of the skull and neck muscles were the animal's main weapon, allowing it to deliver flesh-tearing bites to any victim.

Food *Carcharodontosaurus*'s teeth were smaller and less curved than those of the tyrannosaurs, but even so these giants would have fed on large prey such as the plant-eating sauropods. The shark-like teeth prompted German palaeontologist Ernst Stromer van Reichenbach to name it after today's Great White Shark (*Carcharodon*).

Where did it live? Originally found in Middle Cretaceous strata in Egypt and named in the 1920s, this giant theropod has recently become better known. The discovery of a 1.5m (5ft) skull in Morocco showed that the animal was as big as the tyrannosaurs.

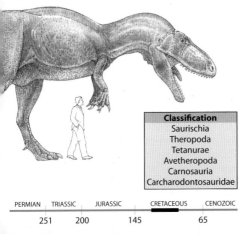

Classification
Saurischia
Theropoda
Tetanurae
Avetheropoda
Carnosauria
Carcharodontosauridae

PERMIAN	TRIASSIC	JURASSIC	CRETACEOUS	CENOZOIC
251	200	145	65	

GIGANOTOSAURUS
JIG-ah-NOH-toh-SORE-us

'Giant southern lizard'

Size Up to 14m (45ft), 5.1 tonnes (5 tons).

Appearance Perhaps the biggest theropod dinosaur, *Giganotosaurus* was similar to *Carcharodontosaurus* (see page 64) and *Tyrannosaurus* (see page 78). As in these giant theropods, the neck was muscular and the head a huge 1.5m (5ft) in length. Consequently, the skull was both light and strong with very large jaw muscles that delivered a powerful bite and had to be flexible enough to absorb the biting stresses.

Giganotosaurus had arms with three-fingered hands, and its whole body was carried by two huge muscular legs and counterbalanced by a massive stiffened tail.

Behaviour Enormous meat eaters such as *Giganotosaurus* may have been scavengers rather than active predators, but recent research shows that their bite was much more powerful than necessary if they were just scavengers.

Food *Giganotosaurus* lived alongside some very large sauropods and may well have hunted down the weakest or youngest of these giants. The discovery in 2000 of part of a *Giganotosaurus* lower jawbone that is 8 per cent larger than those of any other theropod has promoted speculation that this was indeed the biggest land-dwelling carnivore of all time.

Where did it live? The fossils found so far have been taken from middle to Late Cretaceous strata in Patagonia, southern Argentina. The description as a single species in 1995 caused a sensation when its finders claimed that it was bigger than *Tyrannosaurus rex*.

Classification
Saurischia
Theropoda
Tetanurae
Avetheropoda
Carnosauria
Carcharodontosauridae

PERMIAN	TRIASSIC	JURASSIC		CRETACEOUS	CENOZOIC
	251	200	145		65

COMPSOGNATHUS

'Elegant jaw'

komp-SOG-na-thus

Size Up to 1m (3ft), 4.5kg (10lb).

Appearance One of the smallest dinosaurs known, this tiny, slender theropod had long legs, a thin, flexible tail, and a long neck and skull but short arms. The hands had three long, clawed fingers. Its teeth were very sharp and pointed but the jaw was not powerful. Like *Sinosauropteryx* (see page 70), it may have been covered in hair-like down.

Behaviour Capable of running and moving very quickly, this was an agile little hunter that could catch equally swift small prey.

Food The remains of a lizard have been found in the stomach of a *Compsognathus* and so it would appear that it ate any small animal it could catch. The prey ranged from small mammals to lizards, and it may even have been able to catch some insects on the wing.

Where did it live? Subtropical coastal lagoons of Late Jurassic times were the favourite haunt of this little dinosaur, which has been found in southern Germany and the south of France. It was first found in the 1850s by quarrymen at the famous Solnhofen lithographic limestone quarries of Bavaria in southern Germany, where *Archaeopteryx* was later found (see page 102). The single *Compsognathus* species was first described in 1861 and only two partial skeletons are known today.

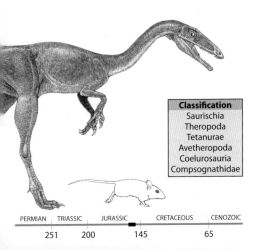

Classification
Saurischia
Theropoda
Tetanurae
Avetheropoda
Coelurosauria
Compsognathidae

PERMIAN	TRIASSIC	JURASSIC	CRETACEOUS	CENOZOIC
251	200	145		65

SINOSAUROPTERYX *'Chinese lizard wing'*
SINE-oh-sore-OP-teriks

Size Up to 1.25m (4ft), 10kg (22lb).

Appearance Although the body shape of *Sinosauropteryx* was similar to other small and agile two-legged dinosaurs, we know that this one had a fluffy covering of down like that of a bird chick or mammal. It had short arms with very flexible wrists and three clawed fingers, one of which was bigger than the others.

Behaviour This close relative of *Compsognathus* (see page 68) used its mobile hands for catching its prey as its jaws were not very strong. Its teeth were sharp and pointed for holding prey animals, which were then probably swallowed whole.

Food Being agile and fast moving, *Sinosauropteryx* ate any small animal it could catch. One specimen has been found with the remains of a small mammal in its stomach while another contains the bones of a lizard.

Where did it live? This dinosaur resided in northeastern China, in subtropical woodlands around lakes that teemed with life ranging from dinosaurs and early birds to fish and insects. It was first found in 1996 within strata of Early Cretaceous age. The hair-like down is preserved as a carbonized film clearly associated with the bones in the rock. The discovery was one of the first of the amazing new finds made in China over the last 20 years.

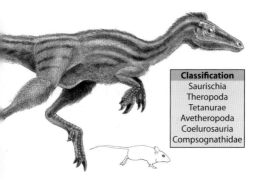

Classification
Saurischia
Theropoda
Tetanurae
Avetheropoda
Coelurosauria
Compsognathidae

PERMIAN	TRIASSIC	JURASSIC	CRETACEOUS	CENOZOIC
251	200	145		65

SCIPIONYX
SHIP-ee-ON-icks

'Scipio-claw'

Size Up to 1.5m (5ft), 25kg (55lb). However, the sole fossil found is of an immature individual only 23cm (9in) long that was perhaps 50cm (20in) when alive and weighed 0.5kg (1lb).

Appearance Just one, near-complete, fossil of *Scipionyx* is known. Only the ends of one hindlimb and the tail are missing. One of the smallest dinosaurs known, it was typical of small two-legged theropods with long legs and a long stiff tail. Being an immature hatchling, its tiny head (5cm/2in long) was big in relation to its body size but had a short snout and large eyes. Most importantly, the fossil preserves soft tissues such as intestine, throat, liver and leg muscles.

Behaviour The position of the liver and intestine is claimed to be more crocodilian than bird-like, giving *Scipionyx* a piston-like mechanism for increasing its lung capacity and activity level without necessarily making it warm-blooded. However, this is speculative, as internal organs can shift as a body decays.

Food The teeth of *Scipionyx* were sharp with serrated edges and its arms had three-fingered hands with sharp grasping claws whose mineralized horny sheaths have been preserved in the fossil remains. However, the hands were rigid and could only grasp by being brought together. Although tiny, this baby was probably adept at catching insects and any small animals it could find.

Where did it live? The skeleton was found in 1981 in Middle Cretaceous lithographic limestone in southern Italy by an amateur collector. *Scipionyx* was described in 1998 and named after the Roman General Scipio, who defeated Hannibal.

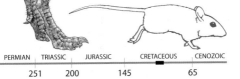

Classification
Saurischia
Theropoda
Tetanurae
Avetheropoda
?Maniraptora

PERMIAN	TRIASSIC	JURASSIC	CRETACEOUS	CENOZOIC
251	200	145		65

ORNITHOLESTES

'Bird robber'

OR-ni-thoh-LES-teez

Size Up to 2m (6ft), 25kg (55lb).

Appearance This primitive coelurosaur was slender with long legs and a long neck, a relatively small skull and a very long tail. The feet and hands each had three prominent toes and fingers, the latter ending in strongly curved claws. The skull was narrow and lightly built but with large nostrils, and the jaws were full of sharp teeth. *Ornitholestes* is thought to be a link in the early

development of birds and may well have been covered with primitive feather-like structures, although there is no fossil evidence for this yet.

Behaviour As its light build and long tail shows, *Ornitholestes* was a fast-running, agile predator. It used its speed and sharp claws to grab small prey animals.

Food Any small animals such as reptiles and mammals would have been potential prey. As its name suggests, it may also have stolen eggs from other reptiles.

Where did it live? The single species was first described by Henry Osborn in 1903 from a partial skeleton found in the early 1900s within Late Jurassic strata of the famous Morrison Formation in western USA. It lived alongside a great diversity of dinosaurs, including the giant predator *Allosaurus* and several kinds of giant plant-eating sauropods such as *Apatosaurus* and *Diplodocus*.

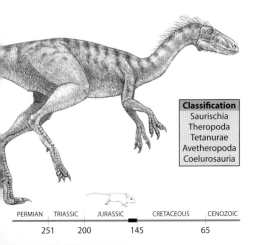

Classification
Saurischia
Theropoda
Tetanurae
Avetheropoda
Coelurosauria

PERMIAN	TRIASSIC	JURASSIC	CRETACEOUS	CENOZOIC
251	200	145	65	

DRYPTOSAURUS
drip-toh-SORE-us

'Tearing lizard'

Size Up to 6m (20ft), 750kg (1,650lb).

Appearance This medium-sized primitive tyrannosaur was one of the first dinosaurs to be discovered in North America, although even now only a small part of its skeleton is known. The fossil bones do show that it had large, powerful legs, a muscular neck, a large head and a long, stiff tail. The jaws were filled with sharp, inwardly curving, blade-shaped teeth.

Although the arms were small they were powerful, and at least one of the three fingers had a large claw.

Behaviour Although *Dryptosaurus* was small compared with other tyrannosaurs, it was probably capable of running quite fast and was the biggest predator in eastern North America.

Food It had plenty of choice of prey among the abundant plant-eating dinosaurs that lived in the region.

Where did it live? *Dryptosaurus* has been found in eastern North America, where in Late Cretaceous times there were extensive low-lying, heavily vegetated coastal plains. Edward Drinker Cope called it *Laelaps* in 1866, but as this name had already been given to an insect, Cope's great rival, Othniel Marsh, renamed it *Dryptosaurus* in 1877.

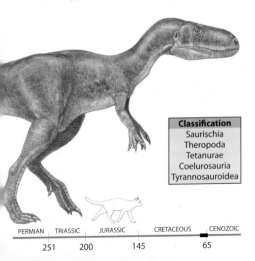

Classification
Saurischia
Theropoda
Tetanurae
Coelurosauria
Tyrannosauroidea

PERMIAN	TRIASSIC	JURASSIC	CRETACEOUS	CENOZOIC
251	200	145		65

TYRANNOSAURUS
'Tyrant lizard'

tie-ran-oh-SORE-us

Size Up to 14m (46ft), 7 tonnes (6.8 tons).

Appearance This was the biggest carnivorous dinosaur until *Giganotosaurus* was found (see page 66). *Tyrannosaurus* was also a giant, with muscular legs that lifted the body over 2m (6.5ft) off the ground. The 1.5m-long (5ft) head and massive neck were counterbalanced by a large stiff muscular tail.

The arms were diminutive in relation to the rest of the body, but were well muscled and the hands had two, clawed fingers.

Behaviour Although lightly built for its size, its skull was well muscled and could deliver a powerful bite. The 20cm-long (8in) curved teeth pierced and gripped the prey, and a twist of its head tore away chunks of flesh. *Tyrannosaurus* was considered a solitary scavenger, but it may have been both ambush hunter and scavenger.

Food Bones of *Edmontosaurus* have been found covered in slash marks probably made by the teeth of *Tyrannosaurus*. And a broken *Triceratops* hip bone suggests that the *Tyrannosaurus's* bite could even crush bone.

Where did it live? The remains of some 30 individuals have now been uncovered in Late Cretaceous strata in western North America. The first reasonably complete skeleton was found in 1902 and it was first described by Henry Osborn in 1905.

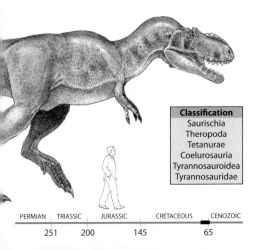

Classification
Saurischia
Theropoda
Tetanurae
Coelurosauria
Tyrannosauroidea
Tyrannosauridae

PERMIAN	TRIASSIC	JURASSIC	CRETACEOUS	CENOZOIC
251	200	145		65

PELECANIMIMUS

'Pelican mimic'

PEL-e-kan-i-MEEM-us

Size Up to 2m (6ft), 25kg (55lb).

Appearance With long legs and a neck that was held upright, this lightly built dinosaur looked a bit like an

Ostrich but for its long arms, three-fingered hands and stiff tail. The head was also birdlike with a pointed 'beaky' skull. The long jaws were lined with 220 very small spiky teeth edged with a few serrations. The rare preservation of skin impressions shows that it had a throat pouch like a pelican, hence the name. A small pointed horn at the back of the skull may have had a skin crest.

Behaviour The shape of *Pelecanimimus*'s large brain indicates that they had good balance for fast running but a poor sense of smell. The head crest and throat pouch may have been highly coloured for display, although the throat pouch may also have been a feeding structure.

Food The unusually small teeth of *Pelecanimimus* are thought to have been used either for filter feeding in the manner of flamingos today, or for holding fish before swallowing them.

Where did it live? A single species is known from a partial skeleton discovered in Early Cretaceous lakeside sediments in Spain, and was first described in 1994.

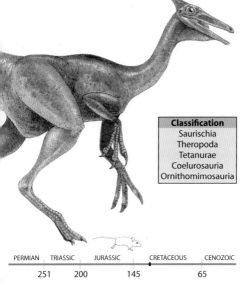

Classification
Saurischia
Theropoda
Tetanurae
Coelurosauria
Ornithomimosauria

PERMIAN	TRIASSIC	JURASSIC	CRETACEOUS	CENOZOIC
251	200	145	65	

ORNITHOMIMUS

'Bird mimic'

OR-ni-thoh-MEEM-us

Size Up to 3.5m (12ft), 70kg (150lb).

Appearance This 'bird mimic' dinosaur had an Ostrich-like appearance with long legs, arms, neck and tail. The small, narrow, wedge-shaped skull had a toothless beak covered in a horny sheath. The brain was large, as were the eyes, which were placed on either side of the skull to give a very wide field of view. The toes and fingers were relatively short and strong with sharp claws.

Behaviour The very long foot, high ankle and relatively long shin indicate that *Ornithomimus* was a fast runner and used its stiff tail as a balance.

Food The toothless beak of *Ornithomimus* suggests that it is most likely to have eaten soft plants, although it may also have caught insects and other small animals, as do many toothless birds today.

Where did it live? It was first described in 1890 by Othniel Marsh from finds discovered in Late Cretaceous rocks in western North America that were originally laid down in coastal plains. Two species are known from a few skulls and incomplete skeletons. Marsh was also first to describe the family group Ornithomimidae, which now contains several genera characterized by short, delicate skulls and long limbs with non-grasping hands.

Classification
Saurischia
Theropoda
Tetanurae
Coelurosauria
Ornithomimosauria
Ornithomimidae

PERMIAN	TRIASSIC	JURASSIC	CRETACEOUS	CENOZOIC
251	200	145	65	

THERIZINOSAURUS   *'Scythe lizard'*
THER-ee-ZINE-oh-SORE-us

Size Up to 11m (36ft), 6 tonnes (5.9 tons).

Appearance This odd dinosaur had the biggest claws known in the animal kingdom. It had immensely long muscular arms (2.5m/8ft) that ended in three-fingered hands with huge 60cm-long (2ft) scythe-shaped claws. No complete skeleton is known so it is unclear what the animal really looked like, but it was probably bipedal with a long neck, small toothless head and horny beak.

Behaviour With such long-clawed hands, this dinosaur must have moved around on its two hind legs. Some experts suggest that it had a long neck and massive hips but short legs and tail, and that it behaved like a Giant Ground Sloth, sitting on its haunches to feed. The arms would have been used to pull down tree branches to eat. Alternatively, it might have used its claws for digging into termite nests.

Food *Therizinosaurus* may have used a horny beak and small leaf-shaped teeth to eat plants.

Where did it live? The first few remains found in Mongolia in 1948 were thought to be those of a sea turtle, as they were found in lake and riverside deposits. But further discoveries described in 1954, from similar Late Cretaceous deposits in Mongolia, show that it was a theropod dinosaur.

Classification
Saurischia
Theropoda
Tetanurae
Coelurosauria
Therizinosauroidea
Therizinosauridae

PERMIAN	TRIASSIC	JURASSIC	CRETACEOUS	CENOZOIC
251	200	145		65

OVIRAPTOR

'Egg thief'

OHVEE-RAP-tor

Size Up to 1.8m (6ft), 23kg (50lb).

Appearance With its long legs and neck, combined with a short tail, *Oviraptor* had an Ostrich-like appearance. Even its short head with large eyes and a toothless but powerful beak was parrot-like. Above the nose rose a prominent bony crest covered in skin

and probably highly coloured for display, similar to that seen in cassowaries today. *Oviraptor* had long, thin arms ending in flexible wrists and three fingers with long, curved claws. It may also have been feathered. The long legs had a high ankle and three long toes.

Behaviour The leg structure shows that this was a fast-moving, agile dinosaur. Its name suggests that it stole eggs but this was a misinterpretation: a recent discovery of an oviraptorid embryo within an egg alongside an adult shows that the parent remained guarding the nest when it was smothered by a sandstorm.

Food *Oviraptor* may have been a meat eater, using its sharp-edged horny beak to chop up small mammals and reptiles caught with its curved claws.

Where did it live? A single skeleton was found in Late Cretaceous strata of Mongolia by Roy Chapman Andrews' expedition from the American Museum of Natural History in New York. *Oviraptor* and all other dinosaurs more closely related to living birds than to *Ornithomimus* are grouped together as maniraptorans.

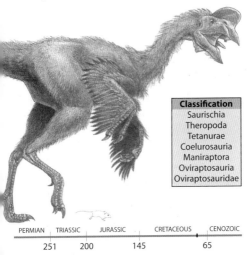

Classification
Saurischia
Theropoda
Tetanurae
Coelurosauria
Maniraptora
Oviraptosauria
Oviraptosauridae

PERMIAN	TRIASSIC	JURASSIC	CRETACEOUS	CENOZOIC
251	200	145		65

TROODON

'Wounding tooth'

TROH-oh-don

Size Up to 1.8m (6ft), 27kg (60lb).

Appearance With its long legs, neck and tail, this dinosaur had a lightly built body. The narrow skull contained a large brain and large, forward-pointing eyes. The jaw was full of small, sharp, inwardly curving teeth with serrations along the inside edge. The thin but muscular arms had flexible wrists, two long clawed fingers and a clawed thumb for grasping its prey. The foot had a retractable second toe for fast running.

Behaviour With their large bird-like brains and binocular vision, *Troodon* were among the most intelligent dinosaurs, capable of hunting down small, fast-moving prey animals. We know from fossil nest sites that, at any one time, up to 24 elongate eggs were laid, narrow end down, in a tight circular pattern. The discovery of adult remains on a nest suggests that they may have been brooded.

Food Troodontids probably preyed on small mammals, dinosaur eggs and hatchlings, as well as insects, and they may have eaten some plant material.

Where did it live? *Troodon* inhabited the fertile coastal plains of the Late Cretaceous seaway of western North America. A single species was described by J. Leidy in 1856, and now some 20 fragments of skulls, teeth and other bones are known, although no complete skeletons have been found.

Classification
Saurischia
Theropoda
Tetanurae
Coelurosauria
Maniraptora
Troodontidae

PERMIAN	TRIASSIC	JURASSIC	CRETACEOUS	CENOZOIC
251	200	145	65	

MEI
MY

'Soundly sleeping'

Size Up to 70cm (28in), 200g (7oz) when adult. However, the sole fossil found is of an immature individual only 53cm (21in) long and weighing 150g (5.3oz).

Appearance This small, primitive troodontid is the first dinosaur found in its natural sleeping position. The skeleton shows it sitting upon its folded legs, with the forelimbs tucked up next to the body. The flexible neck is curved back to the left, with the small head tucked between the left elbow and the body in the manner seen in many sleeping birds today. *Mei* had a small skull with large eyes and nostrils and a relatively large brain. The incompletely fused bones of the skeleton show that the individual was not yet adult.

Behaviour The very bird-like posture displayed by the skeleton helps maintain body temperature and lends support to the claim that some advanced non-avian dinosaurs were warm-blooded. It also indicates an evolutionary connection between maniraptoran dinosaurs and birds.

Food Like other troodontids, *Mei* had beak-like jaws that were lined with numerous, closely packed, small, sharp but not serrated teeth that were well adapted for catching, killing and eating animal food. *Mei* was an active hunter with good eyesight and perhaps a good sense of smell.

Where did it live? *Mei* is yet another of the astonishing dinosaurs found in Early Cretaceous age strata in Liaoning Province, China. Around 130 million years ago this region was forested and dotted with numerous lakes, rivers and active volcanoes. The fossil was described in 2005 by Mark Norell, from America, and Xu Xing, from China.

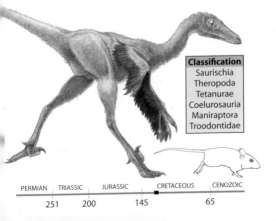

Classification
Saurischia
Theropoda
Tetanurae
Coelurosauria
Maniraptora
Troodontidae

PERMIAN	TRIASSIC	JURASSIC		CRETACEOUS	CENOZOIC
	251	200	145		65

BAMBIRAPTOR
BAM-bee-RAP-tor

'Baby robber'

Size 1.25m (4ft), 3kg (6.6lb) when adult. However, the most complete fossil found is of an immature individual that was only about 70 per cent of adult size, measuring around 1m (3ft) long and 2kg (4.4lb).

Appearance The remains of this small bird-like velociraptorine dromaeosaur are of an immature individual (*Bambiraptor*'s name comes from the Italian *bambino*, meaning 'baby'), so its braincase and eyes were proportionally larger (the skull was 125mm/5in long) than those of an adult. The arms and hands were proportionally longer than those of any other non-avian dinosaur, while the hind legs were like those of living running birds. Otherwise, *Bambiraptor* was a typical maniraptoran carnivore with two functional toes and a large and strongly curved raptorial claw on its third toe.

Behaviour *Bambiraptor*'s light build, long limbs and raptorial claws show that it was an agile predator. Its brain was one of the largest and most advanced relative to body size seen in the dinosaurs. It probably had well-coordinated bird-like sight and movement, but its sense of smell was poorer than that of *Troodon*, the tyrannosaurids and other theropods. It may have lived in trees or have hunted very small agile animals. The range of its arm movement was considerable and similar to the flight stroke of modern birds.

Food The immature specimen found had nine curved, serrated blade-like teeth, but there may have been as many as 12 when the animal was adult.

Where did it live? A nearly complete skull and skeleton were found in the late 1990s in Late Cretaceous strata in Montana, USA, and the species was described in 2000.

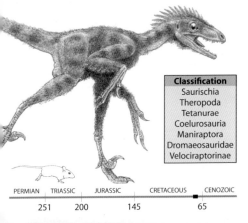

Classification
Saurischia
Theropoda
Tetanurae
Coelurosauria
Maniraptora
Dromaeosauridae
Velociraptorinae

PERMIAN	TRIASSIC	JURASSIC	CRETACEOUS	CENOZOIC
251	200	145	65	

MICRORAPTOR

'Small thief'

MY-cro-RAP-tor

Size 77cm (30in), 200g (7oz).

Appearance This is the smallest dromaeosaurid known, with a body that was only 47mm (19in) long. Like living birds, its body was covered in downy feathers, while the long, stiff, bony tail ended in a tuft of longer feathers. The long arms had hands covered with a dozen flight-like primary feathers and 18 shorter secondary feathers, and the hind limbs had a similar arrangement.

Behaviour From the arrangement of the feathers on both front and hind limbs it is thought that *Microraptor* lived in trees, from which it could launch itself and glide to escape predators or search for food. It has also been claimed that the presence of feathers supports the idea that flight originated from gliding rather than ground-based flapping and take-off. However, flapping flight had already evolved in Late Jurassic times long before the existence of *Microraptor*.

Food Being so small and having the teeth of a meat eater, even if they were serrated only at the tip,

Microraptor probably lived on insects. The serrations would have helped crunch through the insects' tough outer case.

Where did it live? Six specimens of this remarkable little feathered dinosaur are known from Early Cretaceous strata in Liaoning Province, China. The first were found in the late 1990s but were part of a specimen that had been reassembled from two different fossils. A second spectacular species with feathered legs was subsequently described in 2003.

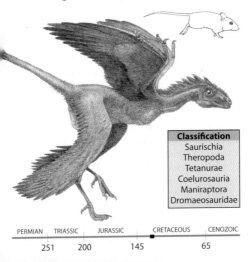

Classification
Saurischia
Theropoda
Tetanurae
Coelurosauria
Maniraptora
Dromaeosauridae

PERMIAN	TRIASSIC	JURASSIC	CRETACEOUS	CENOZOIC
251	200	145	65	

SINORNITHOSAURUS
SINE-or-NITH-o-sore-us

'Chinese bird lizard'

Size 1.2m (4ft), 7kg (15lb), arm span of 0.9m (3ft).

Appearance This is the earliest dromaeosaur known. This little bipedal dinosaur had long limbs and a long tail, stiffened with bony projections of the vertebrae. The body was slender and a lightly built skull housed large eyes. It had a large breastbone and the longest

arms of any dromaeosaur.

Long bristle-like structures stuck out from the body and although flight-type feathers have not yet been recovered from the arms, it is quite likely that they were present. The arm and wrist joints allowed the arm to unfold into a wing-like structure.

Behaviour *Sinornithosaurus*'s long arms ended in three fingers with sharp curved claws. The second or middle finger was by far the longest. The toes of the feet were even more unusual, with the first clawed toe being significantly smaller than the other three. The second toe had the biggest claw, which was retractable, and the remaining two clawed toes were those normally in contact with the ground. This arrangement was perhaps an adaptation for climbing trees and catching prey.

Food The jaws were set with 15 pairs of small, curved, sharp-bladed teeth with serrated edges for eating insects and other small animals.

Where did it live? A nearly complete skull and part of the skeleton of a single specimen with associated feathers were found in the late 1990s in Early Cretaceous strata of Liaoning Province, China. The single species was described in 1999.

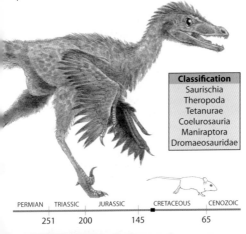

Classification
Saurischia
Theropoda
Tetanurae
Coelurosauria
Maniraptora
Dromaeosauridae

PERMIAN	TRIASSIC	JURASSIC		CRETACEOUS	CENOZOIC
	251	200	145		65

VELOCIRAPTOR *'Quick robber'*
vel-OSS-i-RAP-tor

Size Up to 1.8m (6ft), 30kg (65lb).

Appearance Although not very big, this bird-like dromaeosaur had all the equipment of a ferocious predator – long muscular legs, and sharply clawed toes and fingers. Its skull was light and strong, containing big eyes, a relatively large brain and a long jaw lined with inwardly curving sharp teeth. The tail was stiffened with bony rods. The tail balanced the body with the hips acting as a pivot and allowing agile movement on its two legs. On the foot, the claws could be lifted to prevent them becoming blunt during walking. And when a leg was kicked forward, the second toe, with its big, sharp, curved claw, could be snapped down like a sickle. The long arms had flexible wrists that could move sideways like those of birds.

Behaviour *Velociraptor*'s hunting abilities were famously portrayed in Stephen Spielberg's film of the Michael Crichton sci-fi novel *Jurassic Park*. It is quite likely that this small bird-like dinosaur was a social animal that cooperated with others in hunting animals larger than itself. Like an Ostrich it could kick its legs with devastating effect, and it could use its toe claws to rip or even disembowel its prey.

Food It probably ate anything it could catch, from small mammals to reptiles bigger than itself.

Where did it live? *Velociraptor* bones were first found in Late Cretaceous rocks in Mongolia by the Roy Chapman Andrews' expedition of the 1920s and were described by Henry Osborn in 1924. Since then more specimens of the same species have been found in northwestern China.

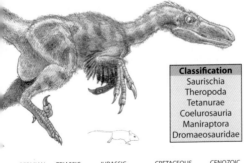

Classification
Saurischia
Theropoda
Tetanurae
Coelurosauria
Maniraptora
Dromaeosauridae

PERMIAN	TRIASSIC	JURASSIC		CRETACEOUS	CENOZOIC
251	200	145		65	

SHUVUUIA

'Bird'

shuh-VOO-yah-ee-ah

Size Up to 1m (3ft) long, 50cm (20in) high and 5kg (12lb).

Appearance This tiny, lightly built theropod had very long hind legs, a long tail, a bird-like neck and skull, and a body covering of fibre-like structures. Tests have shown that these 70-million-year-old fossil fibres still contain a protein called beta-keratin, which

is a characteristic constituent of feathers, and they may therefore be the remains of feather-like structures (*shuvuu* is Mongolian for bird, hence the dinosaur's name). The arms were very strange, in that they were like short, stubby wings ending in a stout claw on the first finger. *Shuvuuia* combines a peculiar mixture of bird-like and non-avialian features.

Behaviour This animal certainly couldn't fly and opinion is divided as to whether it was a peculiar kind of flightless running bird or evolved from some other non-avialian theropods.

Food What exactly *Shuvuuia*'s hand claw was for is also unknown, but it may have been used for digging or for prising bark from trees to get at insects and their larvae. The teeth were small and restricted to the front of the jaws as the small prey was swallowed whole once it was caught. A reduction and subsequent complete loss of teeth is seen in the birds.

Where did it live? Originally, the remains of a single individual were found in the late 1990s in Late Cretaceous strata of Mongolia but more material is now known. A single species was described in 1998.

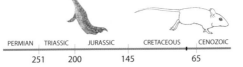

Classification
Saurischia
Theropoda
Tetanurae
Coelurosauria
Maniraptora
Alvarezsauridae

PERMIAN	TRIASSIC	JURASSIC	CRETACEOUS	CENOZOIC
251	200	145		65

ARCHAEOPTERYX   *'Ancient wing'*
ar-kee-OP-ter-icks

Size Up to 50cm (20in), 500g (1lb).

Appearance *Archaeopteryx* resembled a rather long-legged, small bird the size of a magpie, except for its long, bony, feathered tail. However, its beak-like jaws were full of small, sharp teeth and three long, clawed fingers were still present on its feathered arms. As could be seen when it was first named 'ancient wing', the body had a mixture of primitive reptilian features and more advanced bird characteristics. Some specimens have been mistaken for a small dinosaur-like *Compsognathus* (see page 68).

Behaviour The feathers of the wings were true asymmetric flight feathers and very similar to those of modern birds. This shows that *Archaeopteryx* was certainly capable of flying, even if not very fast. It probably used its hand claws to climb up into bushes and trees.

Food It would have eaten anything that moved and could be caught and swallowed whole, from insects to small reptiles.

Where did it live? *Archaeopteryx* was first found in 1861 at Solnhofen within lithographic limestone deposited in coastal subtropical lagoons of Late Jurassic times in Bavaria, southern Germany. It was described by H. von Meyer and a second species was subsequently recognized in 1993. Altogether 10 specimens are now known.

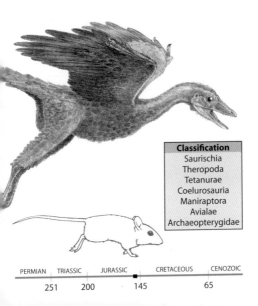

Classification
Saurischia
Theropoda
Tetanurae
Coelurosauria
Maniraptora
Avialae
Archaeopterygidae

PERMIAN	TRIASSIC	JURASSIC	CRETACEOUS	CENOZOIC
251	200	145	65	

ANCHISAURUS

'Near lizard'

AN-kee-SORE-us

Size Up to 2.4m (8ft), 55kg (120lb).

Appearance Like most prosauropods, *Anchisaurus* had a very small head with a proportionally long neck and tail, powerful hind legs and shorter front legs. The thumb claw was much bigger than those of the other fingers and perhaps could be used as a weapon. The form of the skull and jaw was more typical of plant eaters than meat eaters, and the teeth were slender and long. There is also evidence that it swallowed stones to help grind plant material in its gizzard.

Behaviour Although *Anchisaurus* spent much of its time on all fours, it could rear up on its hind legs when necessary to reach for plant food or defend itself with its large thumb claw.

Food *Anchisaurus*'s diet consisted mostly of plant material but it is possible that it also ate some small animals.

Where did it live? The discovery of a nearly complete skull and skeleton were the first dinosaur remains to be found in the Early Jurassic strata of northeastern America (in New England), in 1818. It was given its current name by Othniel Marsh, but not until 1885.

Classification
Saurischia
Sauropodomorpha
Prosauropoda
Anchisauridae

PERMIAN	TRIASSIC	JURASSIC		CRETACEOUS	CENOZOIC
	251	200	145		65

MUSSAURUS

'Mouse lizard'

muss-SORE-us

Size Adults up to 3m (10ft), up to 180kg (400lb).

Appearance The remarkable fossils of this tiny prosauropod include some virtually complete skeletons that are so small they can easily fit in the palm of your hand. Others are 25–30cm (10–12in) long, and all represent hatchlings. More fragmentary remains link these babies to larger individuals, which were probably the adults. Although *Mussaurus* has been linked with *Plateosaurus*, this diagnosis is incorrect according to sauropod experts Peter Galton and Paul Upchurch. The adult remains have yet to be described.

Behaviour Some prosauropods, such as *Anchisaurus* (see page 104), are known only as isolated individuals, while others such as *Mussaurus* and *Plateosaurus* are found in associations of the same species, showing that they habitually lived close to one another, especially when breeding. *Mussaurus* eggs have also been found, although the clutch size is not known (it might have been as low as two eggs). The eggs and hatchlings were clearly very small compared with the adults.

Food The bone structure of *Mussaurus* skeletons shows that they were fast-growing animals. It is possible that, although basically plant eaters, they also ate more energy-rich food such as animal protein.

Where did it live? In 1979 four skulls and the remains of more than ten skeletons found in Late Triassic strata in Argentina, South America, were described as belonging to a single species of this genus. The skeletons ranged from fragmentary to complete and from juvenile to adult.

life size

> **Classification**
> Saurischia
> Sauropodomorpha
> Prosauropoda

PERMIAN	TRIASSIC	JURASSIC	CRETACEOUS	CENOZOIC
251	200	145		65

PLATEOSAURUS

'Flat lizard'

PLAT-ee-oh-SORE-us

Size Up to 9m (30ft), 4 tonnes (3.9 tons).

Appearance Although this was a medium-sized dinosaur in comparison with some of the later sauropods, in Late Triassic times it was one of the largest dinosaurs. Like most prosauropods, *Plateosaurus* had a barrel-shaped body, a massive tail and strong limbs to support them. The hind limbs were longer than the front ones, but the hands were broad and could probably support the body on all fours. The neck was long and the head small, and the jaws were full of small ridged teeth. There were claws on all fingers and toes, these being especially big on the thumbs.

Behaviour Although *Plateosaurus* may sometimes have walked on all fours, with its long hind limbs and shorter front ones, it may also have used the hind ones alone for walking upright. The prominent claws may have been used for defence and for digging up food or catching prey.

Food The teeth show a combination of features that could have been used for eating both plants and meat. Despite this, *Plateosaurus* was likely to have been primarily a herbivore.

Where did it live? *Plateosaurus* was named by the German palaeontologist H. von Meyer in 1837, and was one of the earliest dinosaurs discovered. Ten skulls and hundreds of bones have been found in Late Triassic strata across Europe, from Switzerland through France and Germany to Greenland. At least two species are known, and some fragmentary remains may represent additional species.

Classification
Saurischia
Sauropodomorpha
Prosauropoda
Plateosauridae

PERMIAN	TRIASSIC	JURASSIC	CRETACEOUS	CENOZOIC
251	200	145	65	

MAMENCHISAURUS
mah-MEN-chi-SORE-us

'Mamenchi [Mamenxi] River lizard'

Size Up to 24m (80ft), 28 tonnes (27.5 tons).

Appearance The most spectacular feature of this large dinosaur was its immensely long 15m (49ft) neck, supported by 17 bones, more than in any other sauropod. These vertebrae were partly hollow to cut down on weight, and they had overlapping and stiffening bony rod-like struts. Details of the skull are not well known but it seems *Mamenchisaurus* had a relatively short snout with strong, vertically arranged teeth.

Behaviour The structure of its neck limited its vertical movement – it may have been restricted to sideways sweeping for feeding and lowering the head to drink.

Food It had an entirely plant-based diet and spent much of the day browsing on the abundant vegetation in the marshy environments where it lived.

Where did it live? *Mamenchisaurus* was first discovered near the Mamenxi River in China, after which it was named. It was described in 1954 by Chung Chien Young, a pioneer in the study of Chinese dinosaurs. There are now six known species of *Mamenchisaurus*, mostly dating from the Late Jurassic and all from China.

Classification
Saurischia
Sauropodomorpha
Eusauropoda

PERMIAN	TRIASSIC	JURASSIC	CRETACEOUS	CENOZOIC
251	200	145	65	

CETIOSAURUS
SEE-tee-oh-SORE-us

'Whale lizard'

Size Up to 18m (59ft), 16 tonnes (15.7 tons).

Appearance The massive body of this large sauropod was carried by four strong, pillar-shaped, elephant-like legs. The neck was long and held straight out rather than upright, and it was counterbalanced by the tail. The backbone had some primitive features. For instance, the individual vertebrae had solid bone cores instead of being hollowed out to reduce their weight, as seen in the more advanced sauropods. Unfortunately, no skull has yet been recovered.

Behaviour As *Cetiosaurus* held its neck out straight, the animal probably fed by sweeping its head from side to side to browse on low-level vegetation. It could, however, lower its head to drink and also raise it periodically.

Food It would have eaten low-growing ferns and cycads.

Where did it live? *Cetiosaurus* was first described in 1841 by Richard Owen from just a few back bones found in Middle Jurassic strata in southern England that were deposited on low-lying coastal plains. Owen thought that it was a giant sea-dwelling crocodile – hence the name – and it was not until a partial skeleton was found in 1870 that it was seen to be a sauropod. Many species have been named, although only one or two are known to be valid.

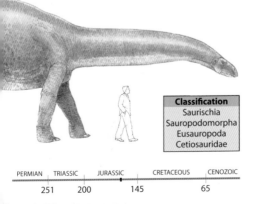

Classification
Saurischia
Sauropodomorpha
Eusauropoda
Cetiosauridae

PERMIAN	TRIASSIC	JURASSIC	CRETACEOUS	CENOZOIC
251	200	145	65	

BARAPASAURUS

'Big-legged lizard'

bah-RAP-a-SORE-us

Size Up to 18m (59ft), 15 tonnes (14.7 tons).

Appearance This was one of the largest of the early sauropods, with long but robust limbs, from which it gets its name. In common with other sauropods, the back bones in the hip region were joined together to form a rigid structure with the pelvis. *Barapasaurus* was one of the first sauropods of gigantic size, and its back bones contained small hollows to help reduce their weight. Unfortunately, no skull *Barapasaurus* has yet been uncovered, but we do know that its teeth expanded into spoon-shaped ends with rough serrated edges for shredding plant food.

Behaviour The large size and weight of *Barapasaurus* meant it habitually walked on all fours and used its long neck to reach its plant food.

Food It was probably a browser that could stretch up into tree canopies and shred leaves from branches with its teeth.

Where did it live? First discovered in 1975, *Barapasaurus* is known from six partial skeletons, all belonging to a single species and all found in southern India's Godavari Valley. It was one of the earliest sauropods and is the best-known Indian dinosaur of Jurassic age. Other fossils found nearby included fish such as coelacanths, crocodiles, pterosaurs and primitive mammals, so it appears *Barapasaurus* lived in coastal swamps where there was plenty of lush plant growth.

Classification
Saurischia
Sauropodomorpha
Eusauropoda

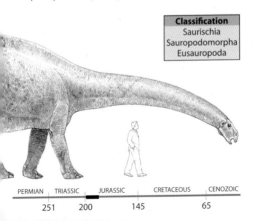

PERMIAN	TRIASSIC	JURASSIC	CRETACEOUS	CENOZOIC
251	200	145		65

AMARGASAURUS *'Lizard from Amarga'*
ah-MAHR-gah-SORE-us

Size Up to 10m (33ft), 4 tonnes (3.9 tons).

Appearance *Amargasaurus* was one of the smaller sauropods and there is no mistaking it. Although it had the usual basic sauropod build, with a small head, heavy body, relatively long neck and tail and four legs, it can easily be distinguished by the highly unusual bony spines along its back. These consisted of two rows of double spines that grew from the individual backbones. The neck was actually quite short by sauropod standards, with just 11 bones, but the neck spines were so

long that they overlapped one another when the neck was raised. There were another 11 back bones with spines and the crest continued over the hips.

Behaviour The spines may have been connected by skin and used for a combination of display and defence, and also perhaps as a heat regulator.

Food *Amargasaurus* had peg-like teeth and was a plant eater.

Where did it live? The single known species was first found 1991 in Early Cretaceous strata at a locality in west-central Argentina, after which it is named. A very similar sauropod, called *Dicraeosaurus*, has been found in Tanzania, Africa.

Classification
Saurischia
Sauropodomorpha
Neosauropoda
Diplodocoidea
Dicraeosauridae

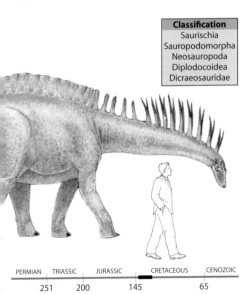

PERMIAN	TRIASSIC	JURASSIC	CRETACEOUS	CENOZOIC
	251	200	145	65

DIPLODOCUS
'Double beam'

di-PLOH-de-kus

Size Up to 27m (88ft), 16 tonnes (15.7 tons).

Appearance This is one of the best known of the giant sauropods, and had a large body, a long neck and long tail made up of 80 vertebrae, and a tiny head. Its weight was supported on strong pillar-shaped elephantine legs, the hind legs longer and more heavily muscled than the front ones. *Diplodocus*'s unusual features included a long tail that was very thin at the end and could have been used as a whip against marauding predators.

Finally, the long peg-shaped teeth were confined to a rake-like structure at the front of the mouth.

Behaviour The structure of the neck and wear on the teeth show that *Diplodocus* could raise its head to reach high plant food. However, since its front legs were shorter than the back ones it may also have fed on shrub-like plants that grew closer to the ground.

Food Wear on the rake-like teeth show that *Diplodocus* stripped leaves from the branches of high-growing conifers and ginkgos, as well as from lower-level ferns and horsetails.

Where did it live? *Diplodocus* was first described in 1884 by Othniel Marsh from a fossil discovered in Late Jurassic strata in western North America. Five nearly complete skeletons of one species and more fragmentary remains of another three are now known.

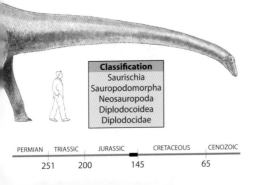

Classification
Saurischia
Sauropodomorpha
Neosauropoda
Diplodocoidea
Diplodocidae

PERMIAN	TRIASSIC	JURASSIC	CRETACEOUS	CENOZOIC
251	200	145		65

APATOSAURUS
ah-PAT-oh-SORE-us

'Deceptive lizard'

Size Up to 24m (80ft), 35 tonnes (34.4 tons).

Appearance This is one of the most famous and familiar dinosaurs because it used to be known as *Brontosaurus*. Like its contemporary *Diplodocus* (see page 118), *Apatosaurus* had a massive body supported by long elephantine legs, a long neck, a tiny head and long whiplash tail. The back bones contained hollows that made them light but strong. As in all sauropods, the bones of the neck and upper part of the back had projections that held ligaments to support the neck and tail. Skulls of these giant sauropods are rarely preserved but it is now known that they were long, low, flat and lightly built.

Behaviour It is possible that despite its size this diplodocid could rear up on its hind legs and use its tail as a back prop in order to reach higher into tree canopies.

Food The teeth at the front of *Apatosaurus*'s mouth were long and peg-shaped, forming a rake-like array. This was very effective in stripping foliage from branches, but the fact that the teeth are often well worn suggests that they were also used for biting off small branches.

Where did it live? Three species are now recognized and were first described by Othniel Marsh in 1877, including what he called *Brontosaurus* in 1879. A total of 11 skeletons have been found, of which two are complete, plus hundreds of separate bones, all of them from Late Jurassic strata in the western USA.

Classification
Saurischia
Sauropodomorpha
Neosauropoda
Diplodocoidea
Diplodocidae

PERMIAN	TRIASSIC	JURASSIC	CRETACEOUS	CENOZOIC
251	200	145	65	

CAMARASAURUS

'Chambered lizard'

kam-AIR-ah-SORE-us

Size Up to 20m (66ft), 20 tonnes (19.6 tons).

Appearance For a sauropod *Camarasaurus* was relatively small, but it is one of the best understood dinosaurs thanks to the many specimens that have been found. It had the classic sauropod form, and although it was not very heavily built its back bones were lightened with hollow spaces – hence its name 'chambered lizard'. Its neck was not very long, with just 12 bones, and these were connected to one another by joints that allowed considerable up-and-down movement of the head. There was, however, less flexibility for side-to-side movement. Its powerful feet had claws, especially on the big toe.

Behaviour This medium-sized plant eater was the most abundant sauropod of its time in North America.

Food With its flexible neck this sauropod could stretch some 8m (26ft) to reach high foliage.

It used its broad snout to push aside branches and grab large amounts of tender leaves. The teeth were broad and strong, and when the jaws closed they locked together, allowing the animal to chew the plant material to some extent before swallowing it.

Where did it live? *Camarasaurus* was first described by Edward Drinker Cope in 1877 from a fossil discovered in Late Jurassic strata in western North America. Today, 20 skeletons are known, of which five or six are nearly complete and can be separated into two species.

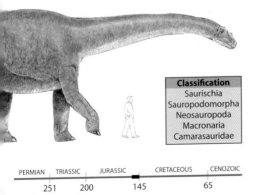

Classification
Saurischia
Sauropodomorpha
Neosauropoda
Macronaria
Camarasauridae

PERMIAN	TRIASSIC	JURASSIC	CRETACEOUS	CENOZOIC
251	200	145	65	

EUHELOPUS

'Good marsh foot'

you-HEEL-uh-pus

Size Up to 15m (49ft), 10 tonnes (9.8 tons).

Appearance This large four-legged plant eater was a typical sauropod dinosaur, its hind legs longer than the front ones, although in *Euhelopus* the front legs were relatively long. It had a massive body with a very long 5m (16ft) neck like that of *Brachiosaurus* (see page 126). As in all dinosaurs, the vertebrae in the tail had bony projections that protected the nerves and blood vessels that ran along the underside of the tail. The head was small and short (only 35cm/14in long), with a steep slope to the snout. The mouth was full of teeth that stuck out forwards.

Each tooth was around 5cm (2in) long with a slightly expanded, flattened and grooved end.

Behaviour The long neck was probably held at an upward angle and swept from side to side when feeding, and then lowered for drinking. It may have been better able to lift its neck vertically than other sauropods such as *Diplodocus* and *Apatosaurus*.

Food It lived solely on plant food and must have spent most of its waking hours feeding on plants such as conifers, ginkgos, cycads, ferns and horsetails.

Where did it live? *Euhelopus* was first discovered in the late 1920s in Late Jurassic strata in China, and only a single species is known from one skull and two incomplete skeletons. The name refers to its large feet, which were thought to have been good for walking on soft ground.

Classification
Saurischia
Sauropodomorpha
Neosauropoda
Macronaria
Titanosauriformes

PERMIAN	TRIASSIC	JURASSIC	CRETACEOUS	CENOZOIC
251	200	145	65	

BRACHIOSAURUS *'Arm lizard'*
BRAK-ee-oh-SORE-us

Size Up to 24m (80ft), 50 tonnes (49 tons), although some experts have claimed it could have weighed as much as 80 tonnes (78.7 tons).

Appearance For a long time *Brachiosaurus* was the largest known dinosaur. It was given its name because it had very long front legs with a high wrist and long toes. The forelegs were bigger than the hind legs and lifted the shoulder some 4.3m (14ft) off the ground. The 12 neck bones, each 70cm (28in) long, extended the neck some 9m (30ft), and like all the back bones they were hollowed to reduce weight.

Behaviour Because the front legs were so long they were very important for walking.

The length of the neck may have created problems in blood flow to the brain. Even giraffes suffer in this way and solve it by having elastic muscular blood vessels; the same solution may have developed in these sauropods.

Food The combined length of the neck and legs lifted *Brachiosaurus*'s head up to 11m (36ft), higher than most other sauropods could reach. Consequently, adults specialized in browsing high-growing trees. The teeth were strong and chisel-shaped.

Where did it live? The first species was described in 1903 from a fossil discovered in Late Jurassic strata in western North America, and a second one was found in 1914 in Tanzania, Africa. Altogether, seven partial skeletons and three skulls are known.

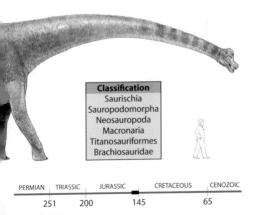

Classification
Saurischia
Sauropodomorpha
Neosauropoda
Macronaria
Titanosauriformes
Brachiosauridae

PERMIAN	TRIASSIC	JURASSIC	CRETACEOUS	CENOZOIC
251	200	145	65	

ARGENTINOSAURUS *'Argentine lizard'*
ar-jen-TEEN-oh-SORE-us

Size It may have been up to 28m (91ft), weighed up to 80 tonnes (78.7 tons), although this is only a guess as not enough is known of the skeleton to be any more definite.

Appearance Using the known back bones as a start point, which are up to 1.2m (4ft) long (about the height of a seven-year-old child), *Argentinosaurus* might have grown to more than 28m (91ft) in length. However, as such it was not the longest dinosaur – at the moment, *Seismosaurus* claims that honour at more than 30m (100ft) in length. Still, *Argentinosaurus* might have been one of the heaviest dinosaurs known, weighing as much as 80 tonnes (78.7 tons).

We can only assume from the few bones preserved that it was similar to other titanosaurs.

Behaviour Once these giants reached maturity they were so big that no predators were a serious threat to them. The sizes they reached approached the maximum possible for any land-based animal.

Food No *Argentinosaurus* skull has been found, so details of its diet remain unknown except that it was certainly a plant eater. Whether it was able to lift its head to feed in tree tops or held its neck out almost straight and browsed by sweeping its head from side to side is also not yet known.

Where did it live? *Argentinosaurus* was first found in 1993 in Cretaceous strata in Argentina, after which it is named.

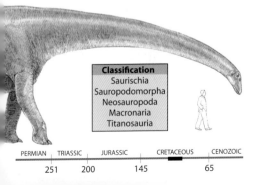

Classification
Saurischia
Sauropodomorpha
Neosauropoda
Macronaria
Titanosauria

PERMIAN	TRIASSIC	JURASSIC	CRETACEOUS	CENOZOIC
251	200	145		65

SALTASAURUS

'Salta lizard'

SAL-ta-sore-us

Size Up 12m (40ft), 25 tonnes (24.6 tons).

Appearance This was a medium-sized titanosaur, but it was probably one of the biggest armoured animals ever to have lived. Its most distinctive feature was its body armour, which consisted of a series of bony studs and knobs, each up to 20cm (8in) wide, that grew embedded in the skin of its back. When the bones of *Saltasaurus* were first found they were thought to belong to an ankylosauran dinosaur because sauropods were not known to possess body armour. Otherwise, *Saltasaurus* had a typical titanosaur body form, although its legs and neck were short compared

with those of many sauropods. The skull is not completely known, but it was probably long and contained long peg-shaped teeth.

Behaviour Although its neck was not particularly long, *Saltasaurus* could still stretch some 6m (20ft) up to reach its plant food. It may also have been able to rear up on its hind legs.

Food Foliage was stripped off branches, and young pine cones and fruit were nipped off using the rake-like array of teeth.

Where did it live? One of the last sauropods to live, *Saltasaurus* was described in 1980 from numerous bones found in Late Cretaceous strata at Salta, Argentina. Later, additional species were added to the genus, based on specimens described in 1929 as species of *Titanosaurus*. The remains of some nine individuals are now known from partial skeletons.

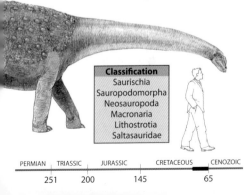

Classification
Saurischia
Sauropodomorpha
Neosauropoda
Macronaria
Lithostrotia
Saltasauridae

PERMIAN	TRIASSIC	JURASSIC		CRETACEOUS	CENOZOIC
251	200		145		65

LESOTHOSAURUS

le-SOO-too-SORE-us

'Lesotho lizard'

Size Up to 1m (3ft), 10kg (20lb).

Appearance *Lesothosaurus* was one of the smallest dinosaurs, and with its lightweight body and bird-like hip bones it was also one of the earliest and most primitive of the ornithischians. It had a short neck, small arms and a long tail, all carried on long,

powerful legs. The legs had high ankles and long feet with four main toes, the third of which was the biggest. Typical ornithischian features in the skull included an extra bone at the front of the jaw and a horny beak. Behind this the jaw had some sharp pointed teeth and then leaf-shaped teeth like those of iguanas today.

Behaviour The structure of the legs and feet show that this small, lightweight dinosaur was capable of running fast to escape predators.

Food Although *Lesothosaurus* was mainly a plant eater, the presence of some sharp-pointed teeth suggests that it might also have eaten small insects and other animals.

Where did it live? This dinosaur is named after the small country of Lesotho in southern Africa, where in 1978 four skulls and associated skeletal material were first found in Early Jurassic strata. Only a single species is known.

The relationships of *Lesothosaurus*, along with the Late Triassic age *Pisanosaurus* from Argentina, to other ornithischians are still unclear.

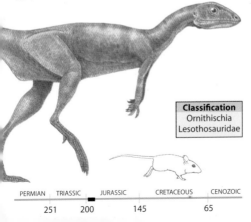

Classification
Ornithischia
Lesothosauridae

PERMIAN	TRIASSIC	JURASSIC	CRETACEOUS	CENOZOIC
251	200	145	65	

SCUTELLOSAURUS   *'Small shield lizard'*
skoo-TELL-o-SORE-us

Size Up to 1.3m (4ft), 20kg (40lb).

Appearance This small ornithischian was a lightly built and agile two-legged dinosaur with a long tail. It was also very unusual in that it was armoured, from which it gets its name. The armour was light compared

with that of its giant descendants such as *Stegosaurus* (see page 140), and consisted of rows of small bony studs embedded in the skin of its back. Altogether there were some 300 studs, which may have been covered with tough horny scales.

Behaviour Although *Scutellosaurus* used its legs for running, it also had quite long arms that may well have been used for walking on all fours. The armour may have helped *Scutellosaurus* deter attack by predators such as *Dilophosaurus* (see page 46).

Food The leaf-shaped teeth were serrated, and since they do not show heavy wear they can have been used only for shredding and chopping quite soft plant food.

Where did it live? *Scutellosaurus* inhabited the hot, dry sand dunes and river valleys in Early Jurassic times in the western USA. A single species was first described in 1981 by Edwin Colbert from a fragmentary skull and skeleton of two individuals.

Classification
Ornithischia
Thyreophora

PERMIAN	TRIASSIC	JURASSIC	CRETACEOUS	CENOZOIC
251	200	145		65

SCELIDOSAURUS

skuh-LID-o-SORE-us

'Hind-leg lizard'

Size Up to 3m (10ft), 340kg (750lb).

Appearance One of the early bird-hipped ornithischians and one of the first armoured dinosaurs, *Scelidosaurus* is very important for our understanding of the history of the thyreophoran dinosaurs. This was a medium-sized ornithischian and had a long and quite heavy body and tail, with relatively short limbs and neck and a small skull. Its neck, back and tail were covered with hundreds of ridged, oval-shaped, bony studs embedded in the skin. The front of the jaws was covered with a

horny beak, behind which were numerous small leaf-shaped teeth.

Behaviour The size of the limbs of *Scelidosaurus* shows that it normally moved around on all fours. It probably could not run very fast and needed its skin armour for protection from predators.

Food *Scelidosaurus* was a herbivore, and may have been able to rise up on its hind legs to browse in high bushes and trees.

Where did it live? *Scelidosaurus* was first found in Early Jurassic strata in southern England and was described by Richard Owen in 1863. Two well-preserved skeletons and partial skulls are known along with some other skeletal remains, all belonging to a single species.

Classification
Ornithischia
Thyreophora
Scelidosauridae

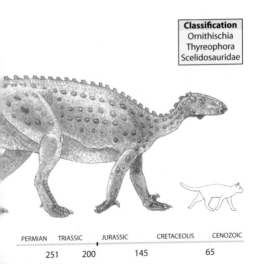

PERMIAN	TRIASSIC	JURASSIC	CRETACEOUS	CENOZOIC
251	200	145		65

HUAYANGOSAURUS *'Huayang lizard'*
hoy-ANG-oh-SORE-us

Size Up to 4m (13ft), 1 tonne (2,200lb).

Appearance *Huayangosaurus* is one of the earliest known and best preserved of the Middle Jurassic stegosaurs. It had short but powerfully built front legs and slightly longer hind legs, a heavy body and a long tail. The head and neck were slung low to the ground. The back and tail had two rows of paired and sharply pointed plates and the tail ended in two more pairs of spikes. The head was quite small and flat, with a strong bony skull. Unusually for a stegosaur, it had teeth at the front of its mouth. The main teeth were rather small, leaf-shaped, and they were set in from the jaw edge.

Behaviour Although *Huayangosaurus* was not very big, like most stegosaurs it was heavily built and probably slow moving, so it depended upon its armour plating and spiked tail for defence against predators.

Food The position of the teeth in the jawbones suggests that *Huayangosaurus* had thin cheeks that helped keep the food in its mouth during chewing. The teeth that have been found are not heavily worn, so perhaps it ate softer plants.

Where did it live? This stegosaur is named after a complete skeleton that was found in 1982 within Middle Jurassic strata in Huayang Province, China. The skull was still attached to the rest of the body, showing that the animal must have been buried quickly. Only a single species is known.

> **Classification**
> Ornithischia
> Thyreophora
> Stegosauria
> Huayangosauridae

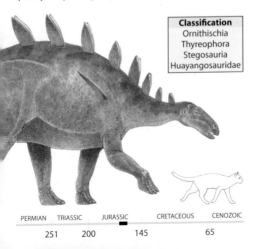

PERMIAN	TRIASSIC	JURASSIC	CRETACEOUS	CENOZOIC
251	200	145		65

STEGOSAURUS

'Roofed lizard'

STEG-oh-SORE-us

Size Up to 9m (30ft), 2 tonnes (1.9 tons).

Appearance This is the most familiar of the stegosaurs, with a large, heavily built body, a small head, a short neck and short front legs. It is easily recognized by the series of large bony plates that ran from the neck down its back to the tail, which ended in two pairs of large bony spikes. The individual plates could be up to 1m (3ft) high and the tail spikes 60cm (2ft) long.

Behaviour
The exact purpose of *Stegosaurus's* plates has been the subject of much debate. The animal was a slow-moving plant eater and vulnerable to attack by large predators. Although the back plates look as if they provided protection, they were not particularly sharp. They seem to have been supplied with numerous blood vessels and so may have acted as heat exchangers for controlling body temperature,

or they may have been used for display. The tail spikes were undoubtedly for protection.

Food *Stegosaurus*'s small teeth were probably used for browsing on cycads and seed ferns.

Where did it live? This was common in western USA during the Late Jurassic; three species have been identified. Altogether, four skeletons are known, of which two are complete, along with a further ten partial skeletons.

Classification
Ornithischia
Thyreophora
Stegosauria
Stegosauridae

PERMIAN	TRIASSIC	JURASSIC	CRETACEOUS	CENOZOIC
251	200	145		65

MINMI
MIN-my

'Minmi [Crossing]'

Size Up to 2.5m (8ft), 700kg (1,500lb).

Appearance This is the most complete and best-known Australian dinosaur, and it was the first armoured dinosaur to be found in the southern hemisphere. Known from a virtually entire skeleton, it is a typical ankylosaur with a broad, squat, heavy body and relatively short legs. The back, upper parts of the legs and flanks were covered with bony stubs and plates (osteoderms) embedded in the skin. There

was also a ridge on the tail composed of pairs of bladed spikes. The head had a broad, flat, solid bony roof and a narrow snout covered with a horny beak at the front.

Behaviour This small ankylosaur was one of the best protected, with an extensive armour covering much of its body surface. Presumably being small and relatively light it risked being overturned by a determined predator.

Food The beak was used for cutting or tearing off large pieces of vegetation, which were then chopped up by rows of small teeth in the cheeks.

Where did it live? *Minmi* was first found in Early Cretaceous strata near the Minmi Crossing (hence its name) in southern Queensland, Australia, in the 1960s. The single known species was further described in 1980 from a second fragmentary skeleton and associated osteoderms. This region lay close to the South Pole at the time, which was forested and although it was not icy it did have dark cool winters.

Classification
Ornithischia
Thyreophora
Eurypoda
Ankylosauria
Ankylosauridae

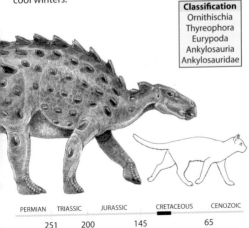

PERMIAN	TRIASSIC	JURASSIC	CRETACEOUS	CENOZOIC
251	200	145		65

EUOPLOCEPHALUS *'Well-protected head'*
you-OH-ploh-SEF-a-lus

Size Up to 6m (20ft), 4 tonnes (3.9 tons).

Appearance This was perhaps the best armoured of
the ankylosaurs and is one of the best known, with
more than a dozen skulls and several good skeletons
now recovered. From head to tail its large, squat,
broad body was protected by rows of bony plates
and spikes, which covered the entire back and parts
of the legs. The stiff, muscular tail ended in a double-
headed club of solid bone. The solid skull roof was
also well protected with a covering of bony plates
and spikes over the neck. Inside the skull was a
convoluted nasal passage whose function is not clear,
although it warmed the air as it was breathed in.

Behaviour
With its broad
and heavy body this ankylosaur relied
on its armour for protection from predators. The tail
club was a very effective weapon that could easily
have smashed the shins of any attacker. The remains of
these dinosaurs are mostly found separate from one
another, suggesting that they normally lived alone.

Food *Euoplocephalus* had a broad horny beak at the front of its mouth for nipping off bunches of plants. The mouth was roofed with a bony palate that allowed the animal to feed and breathe at the same time. The two rows of small leaf-shaped teeth in specimens found are worn only at the tips, showing that *Euoplocephalus* did not chew the plant matter very much but probably had a long tongue that rolled the food up before it swallowed it.

Where did it live? *Euoplocephalus* was first found in Late Cretaceous strata in western North America and a single species was described in the early 1900s.

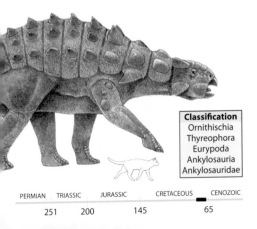

| **Classification** |
| Ornithischia |
| Thyreophora |
| Eurypoda |
| Ankylosauria |
| Ankylosauridae |

PERMIAN	TRIASSIC	JURASSIC	CRETACEOUS	CENOZOIC
251	200	145	65	

HETERODONTOSAURUS
HET-er-oh-DON-toh-SORE-us

'Different-teeth lizard'

Size Up to 1.2m (4ft), 15kg (33lb).

Appearance *Heterodontosaurus* was one of the smallest and earliest of the bird-hipped (ornithischian) dinosaurs, and was a lightly built and fast-moving plant eater. It had a long, stiff tail, a short, curved neck and a small head, along with long hind limbs and quite long and strong arms. Its most distinctive feature was its teeth, which varied in size and shape and gave it its name. It also had cheeks that held plant food in its mouth while it was being chewed.

Behaviour The arm had a flexible wrist, and it had five long fingers and an opposable thumb that could have been used for grasping plant stems. The length of its arms meant that they were often used for grasping. It could escape predators by running quickly on its two hind legs, which had long shins and four long toes.

Food *Heterodontosaurus* had three kinds of teeth – small chopping teeth at the front of the mouth, with bigger grinders at the back and two pairs of large tusks in between, like the canines of mammalian dogs and cats. These differences allowed its usual plant food to be chopped and chewed before being swallowed. The canine-like teeth may have been present only in males, as in some deer living today, and so just used for display.

Where did it live? A single species was first described in 1962 from two skulls and a complete skeleton found in Early Jurassic strata in South Africa. The sediments were originally deposited in landscapes with sand dunes and rivers that dried up seasonally.

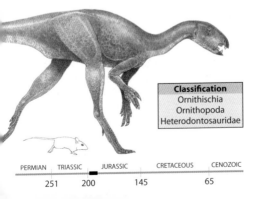

Classification
Ornithischia
Ornithopoda
Heterodontosauridae

PERMIAN	TRIASSIC	JURASSIC	CRETACEOUS	CENOZOIC
251	200	145		65

HYPSILOPHODON
'High ridge tooth'

HIP-si-LOH-foh-don

Size Up to 2m (6ft), 25kg (55lb).

Appearance *Hypsilophodon* was a typical small ornithopod. It had long hind limbs and a long, stiff tail that balanced the weight of its body, neck and head across its hips. The legs had long shins and four long toes, while the shorter arms had five long fingers. The head was the size of a cat's and had large eyes. The front of the mouth had a horny beak and some peg-like teeth in the upper jaw. Behind this the cheek region had a battery of grooved and high-ridged teeth, from which the animal gets its name.

Behaviour *Hypsilophodon* was first described as an agile and rather bird-like animal with feet that allowed it to climb trees. However, it was then realized that the feet were actually those of a fast-running ornithopod.

Food The horny beak was used for nipping off pieces of plant and perhaps shredding them with the front peg-shaped teeth. The strong cheek teeth show that *Hypsilophodon* ate tough plant matter that needed considerable grinding before it was swallowed.

Where did it live? *Hypsilophodon* was one of the earliest dinosaur discoveries – in 1849 in Early Cretaceous strata in southern England – and was at first thought to be a young *Iguanodon*. But then Thomas Huxley, an evolutionist and friend of Charles Darwin, described it as a new kind of dinosaur in 1869. It has since been found in Spain and possibly North America, and three nearly complete skeletons plus another ten partial skeletons are now known.

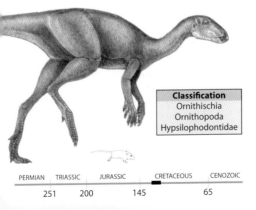

| **Classification** |
| Ornithischia |
| Ornithopoda |
| Hypsilophodontidae |

PERMIAN	TRIASSIC	JURASSIC		CRETACEOUS	CENOZOIC
251	200	145			65

THESCELOSAURUS   *'Wonderful lizard'*
THESS-ul-luh-SORE-us

Size Up to 3.5m (12ft), 30kg (66lb).

Appearance The body, curved neck and head of this medium-sized hypsilophodont were counter-balanced by a long, stiff tail and held up by long, powerful hind legs with four long toes and a longish shin. The arms ended in five-fingered hands and there were small hoof-like claws at the ends of the toes. Rows of bony studs were embedded in the skin of the back but offered little protection from predators. The skull had large holes for the eyes, which suggests that sight was important to the animal.

Behaviour The length of *Thescelosaurus*'s shin in relation to its thigh was not very great, suggesting that it was probably not a very fast runner, although it may have been quick enough to escape most large predators.

Food *Thescelosaurus* mainly foraged for low ground-covering plants and shrubby bushes.

Where did it live? Although this is one of the most primitive ornithopods, its remains were first described in 1913 from Late Cretaceous strata in western North America. In 2000 it was claimed that a skeleton of *Thescelosaurus* found in South Dakota preserved the fossilized remains of the animal's heart but not all experts agree with this diagnosis. The fossils of eight partial skeletons have been recovered to date.

Classification
Ornithischia
Ornithopoda
Hypsilophodontidae

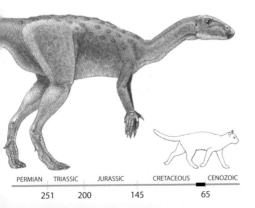

PERMIAN	TRIASSIC	JURASSIC	CRETACEOUS	CENOZOIC
251	200	145	65	

DRYOSAURUS

'Tree lizard'

DRY-oh-SORE-us

Size Up to 3.5m (10ft), 40kg (88lb).

Appearance Thanks to abundant fossil remains, this small ornithopod is one of the best-known dinosaurs. Reconstruction of the skeleton shows that when it was running this small, agile ornithopod held its body, neck and long, stiff tail horizontal. The tail was over half the length of the body and acted as a counterbalance,

and the whole weight of the animal was carried by the long muscular legs and the hips.

Behaviour *Dryosaurus*'s long shins and toes show that it could run fast, and it has been estimated that a small, light ornithopod such as this could reach speeds of up to 40kph (25mph), allowing it to outrun most predators. The large eyes show that keen vision was important for its way of life. Analysis of its bone structure tells us that it grew quickly and continuously from baby to adulthood.

Food *Dryosaurus*'s high, ridged cheek teeth were self-sharpening and very effective at chopping and chewing its plant food.

Where did it live? The Yale University dinosaur expert Othniel Marsh first described *Dryosaurus* in 1894 and named it so because he thought it lived in woodlands. It was one of the most common and widespread dinosaurs of Late Jurassic times, and has been found in western North America, Tanzania in Africa, and in England. The North American material includes one nearly complete skeleton plus seven partial ones. A second species from Tanzania is known from a large number of dissociated bones.

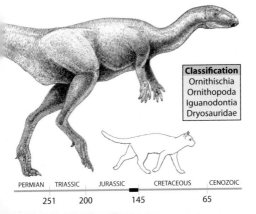

Classification
Ornithischia
Ornithopoda
Iguanodontia
Dryosauridae

PERMIAN	TRIASSIC	JURASSIC	CRETACEOUS	CENOZOIC
251	200	145	65	

CAMPTOSAURUS

'Bent lizard'

KAMP-toh-SORE-us

Size Up to 6m (20ft), 4 tonnes (3.9 tons).

Appearance *Camptosaurus* was a medium-sized ornithopod and being so big it needed a large stomach to digest all its plant food. Otherwise, it had the usual body form of these dinosaurs. The head was long and a bit like that of a horse but with a horny toothless beak at the front of the mouth. The back tendons, especially those of the tail, turned to bone as the animal grew older, stiffening the lower back and tail.

Behaviour *Camptosaurus*'s long, sturdy arms could be used for walking on all fours, although it often walked on its hind legs only. The hand and wrist were also well developed to bear weight, but could equally be used to grasp plants.

Food The battery of tightly packed cheek teeth had rough spiky edges that helped cut and chew coarse plant material.

Where did it live? *Camptosaurus* has been found in Late Jurassic and Early Cretaceous strata in the USA and southern England, showing that the two continents were still connected at this time. When it was first named in 1885, palaeontologists thought that it could bend its back at the hip, and although this is now known to be wrong its name 'bent lizard' is still used. Two or three species are now recognized, and although only one has been found in North America it is known from abundant fossil material, including some 30 skull bones and ten partial skeletons that range from juveniles to adults.

> **Classification**
> Ornithischia
> Ornithopoda
> Iguanodontia
> Camptosauridae

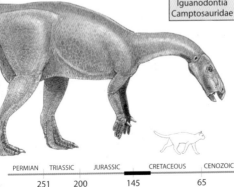

PERMIAN	TRIASSIC	JURASSIC	CRETACEOUS	CENOZOIC
251	200	145	65	

IGUANODON

ig-WAH-noh-don

'Iguana tooth'

Size Up to 10m (33ft), 5 tonnes (4.9 tons).

Appearance *Iguanodon* was one of the larger ornithopods, the reconstruction of which has changed considerably since first attempted by Gideon Mantell in the 1820s. He saw it as a huge, sprawling, lizard-like reptile. Now we know it had the typical ornithopod form and posture. The tail was stiff and held horizontally to counterbalance the heavy body, which was carried on strong hind legs. The arms were long and strong, the hands with three long middle fingers ending in hoofs and the thumb ending in a big spike that Mantell thought was a nose horn. The spike was probably used for defence.

Behaviour The powerful legs ended in large feet with three enormous splayed toes like those of a giant bird. Each long toe ended in a blunt claw and it appears that *Iguanodon* walked on the tips of its toes but supported by a large heel pad, something that is confirmed by their fossil footprints.

Food This dinosaur's name means 'iguana tooth' because its leaf-shaped teeth look like those of the plant-eating iguanas living today. The large, horny beak was used for cutting off foliage that was then further chopped and ground by rows of cheek teeth.

Where did it live? *Iguanodon* remains were found in southern England in the 1820s and it was one of the first known dinosaurs, described by Gideon Mantell in 1825. Remains of six different species have since been found elsewhere in Europe, North America and Mongolia. There are nearly 26 associated skeletons and skulls of one species known from Bernissart in Belgium.

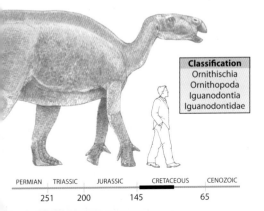

Classification
Ornithischia
Ornithopoda
Iguanodontia
Iguanodontidae

PERMIAN	TRIASSIC	JURASSIC	CRETACEOUS	CENOZOIC
251	200	145		65

OURANOSAURUS *'Brave [monitor] lizard'*
oo-RAN-oh-SORE-us

Size Up to 7m (23ft), 2 tonnes (1.9 tons).

Appearance This medium-sized iguanodont is distinguished from others in its group by some distinctive features. Most prominent of these was the long 'sail' that extended down its back to the tail. Otherwise, it had the usual form, with a long body carried by massive hind limbs that were used for walking. The arms were rather short but it could rest on all fours when necessary. The skull was long with a flat, broad snout covered in a horny beak.

Behaviour The sail was supported by bony projections from the backbone that were well supplied with blood vessels and covered with skin.

It is thought that such structures could have worked as heat exchangers – alternating as solar panels for heating up and as radiators for cooling down. Sails are also seen in the theropod *Spinosaurus* (see page 58), and the isolated plates of *Stegosaurus* (see page

140) may have been similar mechanisms. They may also have been used for fat storage or for display.

Food The tall, ridged cheek teeth allowed *Ouranosaurus* to chew coarse plant material.

Where did it live? A skull and two partial skeletons were found in 1966 and first described in 1976. Only a single species is known of this Early Cretaceous iguanodont, which lived in Niger, Africa, alongside the theropod *Suchomimus*.

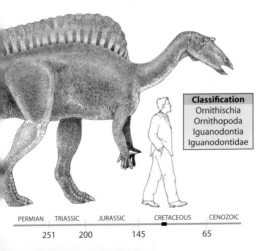

Classification
Ornithischia
Ornithopoda
Iguanodontia
Iguanodontidae

PERMIAN	TRIASSIC	JURASSIC	CRETACEOUS	CENOZOIC
251	200	145		65

HADROSAURUS
'Bulky lizard'

HAD-roh-SORE-us

Size Up to 10m (33ft), 3 tonnes (2.9 tons).

Appearance Historically of great importance, *Hadrosaurus* was the first of the duck-billed dinosaurs found and was described in 1858 by the American anatomist Joseph Leidy. Its name might seem unimaginative, but when it was found it was one of the biggest dinosaurs known. Its *Iguanodon*-like skeleton was the first to be displayed publicly in the USA and was the first nearly complete skeleton discovered. The presence of hind limbs that were much bigger than the front ones showed for the first time that even big dinosaurs could move around just on their two hind legs. Since this posture was unlike that of most living reptiles and unlike the way that dinosaurs had previously been reconstructed, it revolutionized the way dinosaurs were seen.

Behaviour Hadrosaur limbs were very similar to those of the iguanodonts but had only four fingers on each hand. It is likely that *Hadrosaurus* mostly moved around on all fours but could run just on its hind legs.

Food The horny beak at the front of the mouth was used for tearing off pieces of plant that were then chopped and chewed by the ridged cheek teeth.

Where did it live? The original skeleton, missing its skull, was found in Late Cretaceous strata in New Jersey, USA. Since few other remains have been found, this is one of the least known hadrosaurs and consists of only one species.

Classification
Ornithischia
Ornithopoda
Hadrosauridae

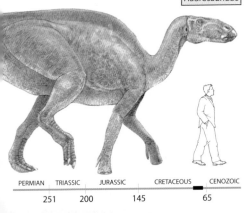

PERMIAN	TRIASSIC	JURASSIC	CRETACEOUS	CENOZOIC
	251	200	145	65

MAIASAURA
MY-uh-SORE-uh

'Good mother lizard'

Size Up to 9m (30ft), 3 tonnes (2.9 tons).

Appearance This medium-sized duck-billed dinosaur had the toothless beak that is so characteristic of the hadrosaurs as well as a distinctive bony knob above the nose. It also had hollows around the nostrils that may have housed inflatable skin pouches for amplifying calls. The large hind legs could support the heavy body, which was counterbalanced by the massively muscular, stiff tail, but normally the animal walked on all fours.

Behaviour This hadrosaur is called 'good mother lizard' because it is one of the dinosaurs for which there is good information about nesting habits and care of young. In 1978, a number of fossilized nests, eggs and babies belonging to *Maiasaura* were discovered at one site called Egg Mountain in Montana, USA. When newly hatched the babies were 50cm (20in) long, but the fossilized skeletons found were 1m (3ft) long and around four weeks old. Studies of the bone structure show that growth was rapid to begin with. Within a year or two *Maiasaura* was more than 3m (10ft) in length, but then growth slowed and it did not reach its adult size until it was about seven or eight years old.

Food The teeth of the babies show signs of wear from eating plant material that must have been supplied to them by the parents.

Where did it live? *Maiasaura* was common in west-central North America in Late Cretaceous times. More than 200 specimens are known, including articulated skulls and skeletons ranging from embryos to adults. The single species was first described by American dinosaur expert Jack Horner in 1979.

> **Classification**
> Ornithischia
> Ornithopoda
> Hadrosauridae

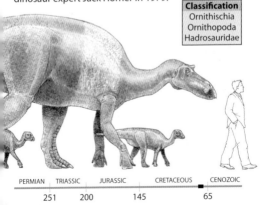

PERMIAN	TRIASSIC	JURASSIC	CRETACEOUS	CENOZOIC
251	200	145		65

LAMBEOSAURUS   *'Lambe's lizard'*
LAM-be-oh-SORE-us

Size Up to 15m (50ft), 7 tonnes (6.8 tons).

Appearance This was one of the largest hadrosaurs known and is distinguished by the strange hatchet-shaped crest on top of its head and a rear-pointing prong at the back. Otherwise, its body shape was very similar to that of other hadrosaurs. The horizontally held massive body was propped up by two powerful hind legs and balanced by a heavy, stiff tail. The front legs were quite long and all the toes had hoof-like claws. The neck was strongly curved and held the head up.

Behaviour We now know that the head crest developed during growth and may have differed between males and females as there is considerable variation. It probably allowed the animals to recognize one another. Like most hadrosaurs *Lambeosaurus* could both walk on all fours and run on its hind legs.

Food The closely packed cheek teeth of hadrosaurs were unusual in that they grew in stacked series, known as a dental battery. There may have been as many as 700 teeth present at any one time, used to grind up plant food.

Where did it live? This dinosaur was first described in 1923 from a find made in Canada in Late Cretaceous strata and was named after the Canadian dinosaur expert L. M. Lambe. Two species are now known from 20 skulls and several articulated parts of their skeletons, gathered from across western North America.

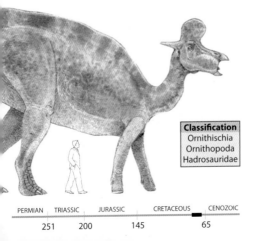

Classification
Ornithischia
Ornithopoda
Hadrosauridae

PERMIAN	TRIASSIC	JURASSIC	CRETACEOUS	CENOZOIC
251	200	145		65

PARASAUROLOPHUS

PAR-a-sore-AH-low-fus

'Beside Saurolophus'

Size Up to 10m (33ft), 5 tonnes (4.9 tons).

Appearance This was one of the larger hadrosaurs, whose most distinctive feature was a curious recurved horn-like structure up to 1m (3ft) long on its skull. There have been many ideas about what this was for. It also had a low ridge running from behind the neck along the back to the powerfully muscled tail. The neck had a strong curve so that the head was held up while the body was horizontal.

Behaviour *Parasaurolophus*'s relatively long front legs show that it could walk on all fours but probably ran on its hind legs. Its head crest was a hollow tube closed at the end. This was not strong enough to have been a weapon and it has been suggested that it was used like a wind instrument – air from the nostrils blown through the tube would have made resonant sounds like a trombone. It is also possible that a web of skin stretched from the tube to the neck and was used for display.

Food The toothless duck bill was used to cut foliage from plants and the cheek teeth were used to chew it well before swallowing.

Where did it live? *Parasaurolophus* was first described in 1922 and is one of the rarest hadrosaurs. Even so, three species are known from two nearly complete skulls and skeletons, along with the remains of three other individuals. The animals lived in Late Cretaceous times in western North America.

Classification
Ornithischia
Ornithopoda
Hadrosauridae

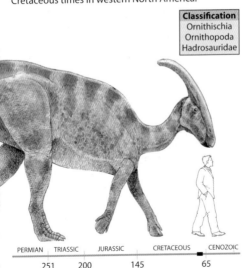

PERMIAN	TRIASSIC	JURASSIC	CRETACEOUS	CENOZOIC
251	200	145	65	

PACHYCEPHALOSAURUS
PAK-ee-SEF-a-loh-Sore-us

'Thick-headed lizard'

Size Because this genus is known from just a few skulls, we can only estimate that it grew to around 8m (26ft) long and weighed up to 2 tonnes (1.9 tons).

Appearance The most distinctive feature of this ornithischian dinosaur was its thick head, as its name suggests. The skull roof had a solid bony dome that could be up to 25cm (10in) thick, and both the snout and back of the skull were covered with bony knobs. Otherwise, we assume that its general shape was like that of the ornithopods, with a bulky body and a long, stiff tail carried on top of long, muscular legs, and

shorter, thinner arms with small hands. The neck was probably thicker than seen in the hadrosaurs.

Behaviour It was thought that the skull dome was used as a battering ram in defence or in head-to-head male disputes over females. However, recent analysis suggests that although the bone was thick it was not particularly strong, and males may just have pushed one another with their heads in trials of strength much as many deer do today. Some *Pachycephalosaurus* skulls were flat and thin, representing different species.

Food The leaf-shaped teeth were small and ridged, indicating that they probably ate relatively tender plant material.

Where did it live? *Pachycephalosaurus* was first described in 1872 under a different name from a fossil found in Late Cretaceous strata in western North America. Its present name was given to it in 1943.

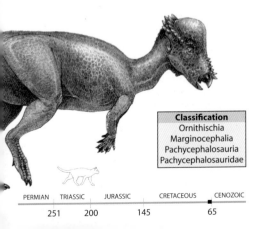

Classification
Ornithischia
Marginocephalia
Pachycephalosauria
Pachycephalosauridae

PERMIAN	TRIASSIC	JURASSIC	CRETACEOUS	CENOZOIC
251	200	145		65

PSITTACOSAURUS

'Parrot lizard'

sit-AK-oh-SORE-us

Size Up to 2m (6ft), 25kg (55lb).

Appearance This small, well-known ceratopsian dinosaur gets its name from its strange parrot-like horny beak. Although *Psittacosaurus* did not look much like the later horned ceratopsians, it is thought to have been closely related to them. The small ledge of bone that stuck out from the back of the skull and the pointed cheekbones were characteristic of the horned ceratopsian dinosaurs. Otherwise, it was lightly built with two long legs with long shins and toes. By contrast, the arms were shorter and more slender, with long fingers that ended in blunt claws. Recent finds have suggested that there might have been long bristle-like structures on the tail.

Behaviour The legs were well adapted for walking, and from the shape of the fingers it seems unlikely that it used its hands for walking. Instead, they were probably used for gathering food.

Food The beak was very good at slicing through plant stems and leaves. Unlike the lower jaws of most dinosaurs, those of *Psittacosaurus* could move back and forth, producing an efficient grinding motion between the upper and lower cheek teeth. Small stones (called gastroliths) swallowed by the animal are commonly found in the stomach cavity, where they helped break down tough plant tissue.

Where did it live? *Psittacosaurus* was first described in 1923 by Henry Osborn from a fossil found in Early Cretaceous strata in Mongolia. Since then a number of skeletons have been found in China, Thailand and central Russia.

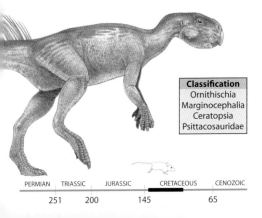

Classification
Ornithischia
Marginocephalia
Ceratopsia
Psittacosauridae

PERMIAN	TRIASSIC	JURASSIC	CRETACEOUS	CENOZOIC
251	200	145		65

PROTOCERATOPS

'First horned face'

PROH-toh-SAIR-uh-tops

Size Up to 2.4m (8ft), 170kg (380lb).

Appearance This small ceratopsian was one of the earliest of the horned dinosaurs, although its name is a bit misleading as it had no real horns, only prominent spiky cheeks and a bony lump on its nose. However, it did have the characteristic frill bones at the back of the skull that projected above the neck and the horny beak of other ceratopsians. The head and thick neck were heavy and carried not far above the ground by the front legs. The hind legs were strongly muscled to lift the broad and heavy body and tail.

Behaviour The discovery of the entangled remains of a *Protoceratops* and a *Velociraptor* suggests that the two died as they fought one another and were buried by a slumping sand dune.

Food The large, powerful, horny parrot-like beak could chop coarse plant stems and perhaps roots, which were then chewed by the battery of grinding teeth in the cheek region.

Where did it live? *Protoceratops* was first discovered in the early 1920s from Late Cretaceous strata in Mongolia by an expedition from the American Museum of Natural History in New York led by Roy Chapman Andrews. It has since been found in China and is known from hundreds of fossil skulls and skeletons.

| **Classification** |
| Ornithischia |
| Marginocephalia |
| Ceratopsia |
| Neoceratopsia |

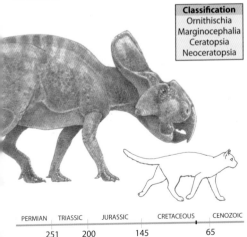

PERMIAN	TRIASSIC	JURASSIC	CRETACEOUS	CENOZOIC
251	200	145		65

PACHYRHINOSAURUS

PAK-ee-RYE-noh-Sore-us

'Thick-nosed lizard'

Size Up to 7m (23ft), 4 tonnes (3.9 tons).

Appearance The most distinctive feature of this large ceratopsian was the rough ridge of bone, more than 18cm (7in) thick, that extended over the length of its snout – hence its name. The ridge may have been covered with skin like the neck frill. The frill itself was adorned with a fringe of spikes and some curved spines. Otherwise, the body was like that of other horned dinosaurs. Both the front and back legs were long and muscular, so that the body, thick neck and massive head (up to 2m/6ft long) were all held horizontal, somewhat like a rhinoceros.

Behaviour The ornamentation of the neck frill was mainly to help individuals recognize members of their own kind and probably to tell the sexes apart. The bony nose ridge was not covered in a horn, as has been claimed, but formed a low, bumpy pad.

Food The mouth's horny beak could cut through thick plant stems that were then chewed by the battery of cheek teeth. These teeth were continuously replaced as they were worn away.

Where did it live? A single species was first described in 1950 from finds in Late Cretaceous strata in western Canada, and it is also known from Alaska. Some 12 partial skulls and a few other fragments of the skeleton are now known.

Classification
Ornithischia
Marginocephalia
Ceratopsia
Neoceratopsia
Ceratopsidae

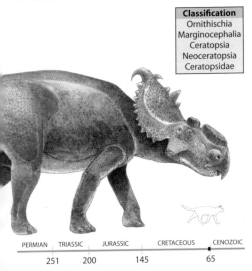

PERMIAN	TRIASSIC	JURASSIC	CRETACEOUS	CENOZOIC
251	200	145		65

TRICERATOPS

try-SAIR-uh-tops

'Three-horned face'

Size Up to 9m (30ft), 6 tonnes (5.9 tons).

Appearance This was the largest and most impressive of the horned dinosaurs. Its name means 'three-horned face' because its skull was adorned with two long horns above the eyes and another shorter nose horn. The head was more than 2m (6ft) long, had a neck frill 1m (3ft) wide and had one of the largest known skulls of any land animal. Inevitably, the legs were powerfully built to support the heavy body and its massive stomach.

Behaviour As *Triceratops* was massive and had ferocious horns, it is unlikely it could have run very fast and in any case it did not need to. However, neck frills have been found with bite marks and punctures from some large theropod predator, so they were certainly used as protection. The frills may also have been used for display, especially between rival males over territory and females. They may also have worked as heat exchangers, helping the animal warm up quickly in the morning and lose excess heat during the day.

Food The huge beak was very good for cutting up tough fibrous plant matter. This was then thoroughly chewed by the batteries of cheek teeth before being swallowed.

Where did it live? *Triceratops* was first described in 1889 by Othniel Marsh from finds made in Late Cretaceous strata in Wyoming, USA. Today, some 50 skulls and partial skeletons of two species are known from western North America, suggesting that it was quite a common dinosaur at that time.

Classification
Ornithischia
Marginocephalia
Ceratopsia
Neoceratopsia
Ceratopsidae

PERMIAN	TRIASSIC	JURASSIC	CRETACEOUS	CENOZOIC
251	200	145		65

GLOSSARY

Allosaurids A family of large bipedal and predatory meat eaters of the Jurassic period belonging within the avetheropod group (Middle Jurassic–Early Cretaceous).

Anchisaurs A group of prosauropods that was first recognized by palaeontologist Othniel Marsh in the 1890s (Early Jurassic).

Ankylosaurs A group of four-legged ornithischian plant eaters armoured with bony plates (Middle Jurassic–Cretaceous).

Archosaurs A major group of reptiles that includes the extinct dinosaurs, pterosaurs and living crocodiles (Early Triassic–Recent).

Avetheropods A group formed from all descendants of the most recent common ancestor of *Allosaurus* and the Aves (Middle Jurassic–Recent).

Aves A group comprising *Archaeopteryx*, living birds and all the descendants of their most recent common ancestor (Late Jurassic–Recent).

Avialians A member of the group Avialae, including all living birds and maniraptorans closer to them than to the dromaeosaur *Deinonychus* (Cretaceous–Recent).

Carnosaurians A group of large meat-eating saurischians comprising *Allosaurus* and all theropods closer to it than to modern birds; sister group to the coelurosaurians (Middle Jurassic–Cretaceous).

Ceratopsians A group of ornithischians, mostly four-legged, with parrot-like beaks and neck frills (Late Jurassic–Cretaceous).

Ceratosaurians The first group of theropods to become widespread and diverse; those theropods more closely related to *Ceratosaurus* than to the birds (Jurassic–Cretaceous).

Cladistics A method of classification that searches for natural groupings of organisms based on shared derived characters.

Coelurosaurians A major group of saurischians that includes living birds and all theropods sharing a more recent common ancestor with the living birds than with *Allosaurus* (Middle Jurassic–Recent).

Conifers A widely distributed and ancient group of seed plants with reproductive cones and needle-like leaves (Carboniferous–Recent).

Cycads A group of tropical seed-bearing plants with palm or fern-like habits (Permian–Recent).

Dinosauromorphs A group that includes all dinosaurs and dinosaur-like archosaurs, such as *Lagerpeton* and *Marasuchus* of Middle Triassic age.

Diplodocids A group of large four-legged sauropods that browsed on relatively tender, low-growing plants (Jurassic–Early Cretaceous).

Dromaeosaurids A group of small to medium-sized meat-eating, bipedal maniraptoran theropods closely related to the avialians, some of which grew feather-like structures (Cretaceous).

Ectotherm An animal that regulates its body heat from external sources such as the sun.

End-Cretaceous extinction event At the end of Cretaceous times, 65 million years ago, the coincident loss of several major reptile groups such as the dinosaurs and pterosaurs, as well as the ammonites and some 60 per cent of all life, marked a global mass extinction.

Endotherm An animal that regulates its body heat from an internal source of energy.

Ferns An ancient group of plants with large feathery leaves that contains more than 12,000 living species (Carboniferous–Recent).

Gastrolith A stomach stone, swallowed by an animal, as an aid to the digestion of tough plant material in the stomach or crop.

Ginkgos A group of primitive plants with fan-shaped leaves; there is only one surviving species (Permian–Recent).

Hadrosaurs A group of large duck-billed, plant-eating ornithopods that generally walked on their hind legs but could go on all fours at times. Some hadrosaurs had ornate head crests (Late Cretaceous).

Horsetails (sphenopsids) An ancient group of plants with jointed hollow stems, circlets of narrow branches and tiny scale-like leaves; only a few species survive today (Devonian–Recent).

Hypsilophodontids A group of small, bipedal, fast-moving, Ostrich-like ornithopods with both horny beaks and cheek teeth for eating plants (Middle Jurassic–Cretaceous).

Ichthyosaurs A group of extinct marine predatory reptiles with streamlined dolphin-like bodies (Early Triassic–Cretaceous).

Iguanodontids A large and varied group of plant-eating ornithopods, including one of the first reptiles to be recognized as a dinosaur (Late Jurassic–Cretaceous).

Maniraptorans A group of small toothless coelurosaurian theropods, such as *Oviraptor* (Jurassic–Cretaceous).

Marginocephalians A group based on the most recent common ancestor of *Pachycephalosaurus* and *Triceratops* and all its descendants (Cretaceous).

Megalosaurs A group of large, heavily built, bipedal carnosaurs with large heads, short strong necks and long tails (Middle Jurassic–Middle Cretaceous).

Ornithischians Bird-hipped dinosaurs, all of which were plant eaters. One of the two main groups of dinosaur, the other being the saurischians (Late Triassic–Late Cretaceous).

Ornithomimids A group of fast-running bipedal theropods, often called Ostrich-dinosaurs because of their long legs and necks. They lacked teeth and had a horny beak (Cretaceous).

Ornithopods Plant-eating ornithischians, mostly bipedal, although some used all four legs from time to time. They had special teeth for eating tough plants (Jurassic–Cretaceous).

Osteoderm A bony stud or plate growing within the skin.

Oviraptorids A group of small maniraptoran theropods with unusual skulls and toothless jaws (Cretaceous).

Pachycephalosaurs A group of large plant-eating ornithischians with thick, domed, bony skulls. (Cretaceous).

Palaeoichnology The study of fossil footprints, trackways and other traces of past life, both plant and animal.

Pangaea An ancient supercontinent formed by all the land masses that came together in Triassic times and split up in Cretaceous times.

Plesiosaurs A group of extinct marine predatory reptiles that typically had four paddle-like flippers, long necks and small heads (Late Triassic–Cretaceous).

Predator An animal that hunts other animals for food.

Prosauropods Long-necked plant eaters that form one of three main groups of saurischians, and that were the first large-bodied dinosaurs. Some were bipedal, but others walked on all fours (Late Triassic–Early Jurassic).

Psittacosaurids A group based on the genus *Psittacosaurus*. Along with the neoceratopsians they form the ceratopsians (Cretaceous).

Pterosaurs An extinct group of flying reptiles, ranging from the size of a sparrow up to that of a small aeroplane. They had wings made from webs of skin supported by an elongate fourth finger (Middle Triassic–Cretaceous).

Quadrupedal Walking on all fours.

Rhynchosaurs An extinct group of squat pig-like reptiles with hooked, parrot-like beaks (Middle Triassic).

Saurischians The lizard-hipped dinosaurs, which varied greatly in size, form and habits. This is one of the two main groups of dinosaur, the other being the ornithischians (Late Triassic–Cretaceous).

Sauropods One of the three main groups of saurischians, which included the biggest dinosaurs and land animals ever known. They walked on all fours, had long necks and tails, and had barrel-shaped bodies (Late Triassic–Late Cretaceous).

Sauropodomorphs A basal grouping of all saurischians more closely related to *Saltasaurus* than to theropods (Late Triassic–Cretaceous).

Seed ferns (pteridosperms) An extinct group of seed-producing plants with fern-like foliage (Late Devonian–Late Cretaceous).

Spinosaurus A group of highly unusual theropods that are among the basal tetanurans (Cretaceous).

Stegosaurs One of five main groups of plant-eating ornithischians that walked on all fours and typically had rows of bony plates and spines along their backs (Middle Jurassic–Cretaceous).

Tetanurans A large group of saurischians that includes all theropods that share a more recent common ancestor with the birds than with the ceratosaurs.

Theropods Bipedal meat eaters that make up 40 per cent of all known dinosaurs. Sister group to sauropodomorphs and one of three main groups of saurischians (Late Triassic–Cretaceous).

Thyreophorans A large group of armoured ornithischians that includes the stegosaurs and ankylosaurs (Jurassic–Cretaceous).

Titanosaurs The last surviving group of plant-eating four-legged sauropods. They were relatively small and had long peg-like teeth (Cretaceous).

Troodontids A group of small, long-legged theropods with many teeth, light skulls and relatively large brains (Cretaceous).

Tyrannosaurids A distinctive group of large, bipedal and carnivorous coelurosaur theropods with large heads and small arms (Late Cretaceous).

Velociraptorines A group of small dromaeosaurids such as *Velociraptor*, distinguished by features of their teeth (Late Cretaceous).

WEBSITES AND FURTHER READING

Websites
http://www.amnh.org
http://www.nmnh.si.edu/paleo/dinosaurs/index.htm
http://www.ucmp.berkeley.edu

Further Reading

Carpenter, K, *Eggs, Nests and Baby Dinosaurs*, 1999, Indiana U. Press.

Carpenter, K (ed.), *The Armored Dinosaurs*, 2001, Indiana U. Press.

Carpenter, K (ed.), *The Carnivorous Dinosaurs*, 2005, Indiana U. Press.

Currie, PJ *et al.* (eds), *Feathered Dragons*, 2004, Indiana U. Press.

Farlow, JO, and Brett-Surman, MK (eds), *The Complete Dinosaur*, 1999, Indiana U. Press.

Tidwell, V, and Carpenter, K (eds), *Thunder-lizards: The sauropodomorph Dinosaurs*, 2005, Indiana U. Press.

Weishampel, DB *et al.* (eds), *The Dinosauria*, 2nd edition, 2004, U. California Press.

Most of these books are fairly academic but readable in part, well illustrated and up-to-date in a rapidly developing science. More general accounts include:

Fastovsky, DE, and Weishampel, DB, *The Evolution and Extinction of the Dinosaurs*, 2nd edition, 2005, Cambridge U. Press.

Martin, AJ, *Introduction to the Study of Dinosaurs*, 2nd edition, 2005, Blackwell Publishing.

INDEX